Studies in Fuzziness and Soft Computing

Volume 405

Series Editor

The series “Studies in Fuzziness and Soft Computing” contains publications on various topics in the area of soft computing, which include fuzzy sets, rough sets, neural networks, evolutionary computation, probabilistic and evidential reasoning, multi-valued logic, and related fields. The publications within “Studies in Fuzziness and Soft Computing” are primarily monographs and edited volumes. They cover significant recent developments in the field, both of a foundational and applicable character. An important feature of the series is its short publication time and world-wide distribution. This permits a rapid and broad dissemination of research results.

Indexed by ISI, DBLP and Ulrichs, SCOPUS, Zentralblatt Math, GeoRef, Current Mathematical Publications, IngentaConnect, MetaPress and Springerlink. The books of the series are submitted for indexing to Web of Science.

More information about this series at http://www.springer.com/series/2941

Ali Guidara

Policy Decision Modeling with Fuzzy Logic

Theoretical and Computational Aspects

With a Case Study: Cuban Missile Crisis

Springer

Ali Guidara
Paris, France

ISSN 1434-9922 ISSN 1860-0808 (electronic)
Studies in Fuzziness and Soft Computing
ISBN 978-3-030-62627-3 ISBN 978-3-030-62628-0 (eBook)
https://doi.org/10.1007/978-3-030-62628-0

This Springer imprint is published by the registered company Springer Nature Switzerland AG
The registered company address is: Gewerbestrasse 11, 6330 Cham, Switzerland

To my three loves,

Odile, Sophian, Amin.

Preface

This book aims to explore the sub-systemic environment of the decision process in public policy with an approach that constitutes the crossroads of several fields, namely policy decision analysis, complex systems, modeling and simulation, as well as fuzzy logic, an artificial intelligence method.

This multidisciplinary approach is at the core of this research project and constitutes a major methodological innovation in public policy studies.

Classical analytical approaches in public policy are limited to the systemic decision-making level. This is due to the absence of adequate tools, even though the sub-systemic level is fundamental to the decision process. This micro-level environment is influenced by several factors and dynamics that make it a complex system, which requires appropriate methodologies such as modeling and simulation.

Modeling is a conceptual representation of a system which involves identifying the components that constitute the system and their dynamics. Simulation requires an appropriate method and an adequate computing platform. However, to model the sub-systemic environment of the policy decision process, a consistent link between this complex system and the field public policy must be established to bridge the two areas and to validate the approach.

This research consists of the development of a new tool to model and simulate the sub-systemic environment of the policy decision process as a complex system. This tool constitutes an innovation based on several theories and techniques. Furthermore, it proposes to enrich the fields of public policy and decision modeling and simulation with the integration of different, yet complementary fields.

Paris, France — Ali Guidara

Synopsis

The topic of public policy decision-making has received considerable attention by scholars which in turn produced several analytical approaches to explain decisionmakers' choices. However, these approaches are limited to the systemic decision-making level.

The purpose of this book is to introduce the policy decision emergence and the dynamics that drive this emergence at the sub systemic level of the decision process. This level constitutes the breeding ground of the emergence of policy decisions but remains unexplored due to the absence of adequate tools.

The sub systemic environment is a nonlinear complex system made of several entities that interact dynamically. The behavior of such a system cannot be predicted or calculated with linear and deterministic methods but needs modeling and simulation to understand and forecast their dynamical evolution.

Simulation requires the development of a model that represents the system. Additionally, to be representative of the policy decision emergence, the requested model must be based on policy decision-making theories.

A tool of complexity, the Stacey Matrix adapted to the public policy field, inspires a link between complexity and public policy through the multiple streams theory. This approach makes it possible to develop a conceptual model made of variables and factors that represents the sub systemic environment of the policy decision process.

An examination of the conceptual model shows that its components are described by vague and uncertain notions that depend on human reasoning. Therefore, it requires an appropriate artificial intelligence method like fuzzy logic to build the computational model of the policy decision emergence and perform the simulation.

The computational model is a multi-level fuzzy inference system that constitutes the policy decision emergence simulation model (PODESIM) developed in this book. PODESIM is an experimental decision diagnostic tool that allows identifying the sub systemic levers of decision emergence using fuzzy data and decision heuristics. It represents an innovation in computational decision-making and a major advancement in the field of public policy.

The multidisciplinary approach developed in this book constitutes the crossroads of several fields, namely decision analysis, complex systems, modeling and simulation, as well as artificial intelligence. It paves the way for policy decision emergence modeling and simulation by bridging complex systems theory, multiple streams theory, and fuzzy logic theory.

Contents

1 Introduction 1
References 5

2 Decision Process and Analytical Frameworks—Levels of Analysis and Paradigmatic Evolution 7
2.1 Decision Analysis Frameworks 8
2.1.1 Rational Decision-Making Model 8
2.1.2 Incremental Model 10
2.1.3 Bureaucratic Politics Model 11
2.1.4 Garbage Can Model 13
2.1.5 Chapter's Conclusion 15
2.2 Decision Emergence as a Complex System 19
References 19

3 Complex Systems and Public Policy 21
3.1 Properties of Complex Systems 24
3.2 Modeling and Simulation of Complex Systems 26
3.2.1 Modeling 26
3.2.2 Simulation 28
3.3 Stacey Matrix: A Complexity Tool 30
References 33

4 Multiple Streams Theory 35
4.1 Basic Foundations of Multiple Streams Theory 35
4.1.1 Problem Stream 36
4.1.2 Policy Stream 37
4.1.3 Politics Stream 38
4.2 Assessment of Multiple-Streams Theory 39
4.3 Paradigmatic and Methodological Transition 42
References 44

5 Artificial Intelligence and Fuzzy Logic . . . 47
5.1 Artificial and Computational Intelligence . . . 47
5.2 Fuzzy Logic . . . 48
5.3 Fuzzy Sets . . . 49
5.4 Fuzzy Logic as a Decision Tool . . . 51
5.5 Fuzzy Inference Systems: Principles and Modelling . . . 53
5.5.1 Structure and Modelling of Fuzzy Inference Systems . . . 53
5.5.2 Fuzzy Inference Algorithms . . . 55
5.6 Mamdani Fuzzy Model . . . 56
5.6.1 Fuzzification of Input Variables . . . 56
5.6.2 Membership Functions . . . 57
5.6.3 Fuzzy Inference Rules and Logical Operators . . . 60
5.6.4 Defuzzification . . . 62
5.6.5 Modeling and Simulation Tools . . . 64
5.7 Conclusion . . . 66
References . . . 67

6 PODESIM—Policy Decision Emergence Simulation Model . . . 69
6.1 Membership Functions of PODESIM . . . 71
6.2 Fuzzy Inference Rules of PODESIM . . . 74
6.3 Case Study—Model Validation and Results . . . 79
6.3.1 Empirical Case Choice . . . 79
6.3.2 Data Collection . . . 80
6.3.3 Simulation and Results . . . 83
6.3.4 Methodological Interpretation of Simulation Results . . . 88
References . . . 89

7 Analysis of Results . . . 91
7.1 Day-by-Day Analysis . . . 93
7.2 Conclusion . . . 112
7.3 PODESIM Limitations and Further Development . . . 112
References . . . 114

8 Innovation and Contributions . . . 115
References . . . 120

9 Conclusion and Future Research . . . 123
References . . . 131

Uncited References . . . 133

Chapter 1
Introduction

We build too many walls and not enough bridges.

Isaac Newton

Decision analysis in public policy has evolved alongside scientific and methodological progress. Paradigms have been developed and analytical models settled attempting to explain and interpret political actions, focusing mainly on the analysis of policy decisions and the explanation of their processes and purposes.

However, the growing complexity of the political, social, economic, and international fields has gradually demonstrated the limits of existing approaches. This fact led to the design of new paradigms intended to better explain policy decisions. These changes brought about by the evolution of the state of the world require a review of the theories and concepts of decision analysis in public policy.

This review aims to preserve concepts that have proved their usefulness and to modify or to reject those that are no longer adequate and thus to identify theoretical gaps to be filled.

Traditional approaches have aimed to interpret state decision-making and to analyze the purpose of the decisions at the systemic or macroscopic level that of the actors and the decision-makers. However, the decision process and its multiple layers, especially the sub-systemic level, are not explored.

The sub-systemic environment of the decision process is that of the components at the deepest level of the decision process, and this level is important for the decision-making analysis and understanding.

According to Mintz and DeRouen (2010), understanding decisions does not provide a complete analysis (p. 10). The process, that generates the decision, is rich in lessons, and it is driven by several entities and dynamics that are at the source of the emergence of a decision and thus policy. Therefore, it must be rigorously investigated. Indeed, Yetiv (2011) argues that the process is a key aspect of decision-making. He states that the process is a vital part of decision-making and deserves explanation (p. 203).

A. Guidara, *Policy Decision Modeling with Fuzzy Logic*, Studies in Fuzziness
and Soft Computing 405, https://doi.org/10.1007/978-3-030-62628-0_1

Traditional analyses that deal with policy decision processes only represent theoretical add-ons to conventional decision-making analysis without addressing the underlying process, although it exists even if a policy decision is not made. The existing processes that may generate decision circumstances do not necessarily result in decision-making. A decision process that does not generate a decision can also be as important and informative as one that results in decision-making, and its consequences can be as rich as a process that results in decision-making.

The increasing complexity of the decision process is at the core of public policy and governance. It is the reflection of the decentralization of powers and influences that affect decision-making; thus, it is essential for the development and implementation of public policies. The determination of the levers of decision emergence and its underlying dynamics is therefore necessary and very useful.

This emergence is the result of a complex process and requires appropriate methods that go beyond the linear techniques provided by analytical approaches which are reductionist and determinist methods. The limitations of traditional linear approaches in decision analysis are indeed mentioned by many authors (Jervis, 1997; Moss & Edmonds, 2005; Lempert, 2002).

Morçöl (2003) raises even the weaknesses of classical statistical tools in policy analysis. He argues that, despite some successes of these tools, their weaknesses have emerged because of the general increase in the complexity of politics and the growing uncertainty that characterizes political issues.

In addition, statistical tools use sets of assumptions and conditions to satisfy the criteria of the searched solution. These tools only contribute to the analysis of data after the studied events have taken place, or during these events.

The complex new context of policy-making requires appropriate methods to address the growing complexity of public policy and especially of the decision process. Some authors advocate for an essential transition toward contemporary policy decision analysis. They plead for a paradigm shift from a positivist approach to a post-positivist methodology that considers the context studied and its specificities (De León & Vogenbeck, 2007). The purpose of this initiative is a complete exploration of the decision process and a good understanding of political actions.

Harrison argues that the influence of authority in the modern world political system can be better captured through complexity concepts than through a simple model that has been relevant in a past era of state dominance (in Harrison 2006, p. 188).

These clarifications about the complexity of policy decisions appeal to a need for approaches that consider this complexity, particularly at the sub-systemic level where the dynamics and interactions are subtle.

This observation prompted us to explore the potential of complexity theory, or complex systems theory, as a means to understand the process that generates decision-making circumstances in public policy. We call this process the decision emergence, and we aim to explore the factors and dynamics that drive this emergence. Complex systems theory challenges long-held views of science [reductionism, linearity] and offers a new set of concepts to understand complex problems (Mitchell, 2009, pp. 300–301).

Th main interest of complex systems theory lies in its capacity to explain emergence phenomena and the dynamics that generate them. Public policies can be considered as emerging phenomena (Morçöl, 2003) according to the concepts of complexity theory because they are complex systems (Morçöl, 2012) and their dynamics constitute the decision-making "black box".

Therefore, we need appropriate methodologies and a multidisciplinary approach because of the heterogeneity of such complex systems and the diversity of their components and dynamics. In other words, we first need to explore the tools developed by complex systems theory to identify the methodology that can deal with the sub-systemic environment of the decision process.

Complexity theory only provides theoretical principles to deal with complex systems, but it does not suggest specific tools for the field of public policy.

Besides the need to identify a useful tool to apply to the decision emergence phenomena as a complex system, it is also essential to find a method that bridges the tool with the public policy field.

The notion of emergence herein corresponds to the meaning determined by complex systems theory, i.e., a rise of behavior at the systemic level as a result of the nonlinear interactions among the sub-systemic components of a system.

Also, a complex system is, in general, a system composed of several components that may have direct or indirect interactions in a decentralized environment. These interactions may generate new systemic behavior which is not predictable from the individual properties of the components.

In the decision process, the systemic behavior represents the emergence of the decision circumstances, called here the decision emergence, which is a preliminary step and an essential condition to decision-making and policy development. This important stage of the decision process requires an exploration of the sub-systemic environment to determine the entities and the dynamics that underlie the decision emergence.

However, this sub-systemic environment is not considered by conventional decision analysis approaches, because of its complex nature and the lack of adequate tools. Therefore, the question is how to identify the components of the sub-systemic environment of the policy decision process and their dynamics as well as how to determine their impact on the decision emergence.

The response to this inquiry requires a rigorous investigation through the following multidisciplinary theoretical and methodological directions.

Chapter 2 introduces the decision process and a review of analytical public policy decision approaches and their paradigmatic evolution. In this chapter, we also examine the level of analysis of the decision process of these approaches and assess the aptitude of the analytical approaches to capture the sub-systemic environment of the policy decision process. Finally, we define the conceptual framework of policy decision emergence and the necessary steps to capture this emergence and its properties.

Chapter 3 deals with complexity theory or complex systems theory and the contributions of this theory to public policy analysis. We establish the conceptual link between complexity theory and policy decision-making through the Stacey Matrix adapted to the public policy field.

We also demonstrate the need for modeling and simulation of the decision process to understand this process and to grasp its dynamics and evolution.

Chapter 4 consists of a review of the multiple streams theory that constitutes the connection between complex systems theory and the decision emergence. We proceed to an assessment of the multiple streams theory that highlights the fundamental characteristics of the policy decision emergence.

We also underline the dynamics of the sub-systemic environment of the decision process and identify its entities, and we demonstrate its uncertain and vague nature. Finally, we define the conceptual model that represents the decision emergence and its components.

This review showed the need for a methodological transition to capture the dynamics of the sub-systemic level of the decision process that are not considered by the analytical approaches. We identified the fuzzy logic theory as the appropriate approach to deal with the vague and uncertain nature of the sub-systemic environment of the decision process.

Fuzzy logic is a method of artificial intelligence and a rule-based methodology. It is an appropriate approach to handle the fuzziness of processes characterized by vagueness, uncertainty, and subjectivity of human reasoning, such as the decision emergence phenomena.

In Chap. 5, we elaborate on the concepts of artificial intelligence, fuzzy logic, and fuzzy sets, and we focus on fuzzy inference systems as an appropriate method to deal with the policy decision emergence model and its sub-systemic structure.

The introduction of this method in the policy decision field constitutes a major advancement and a methodological transition toward the effective integration of computational methodologies in this field.

We then approach the modeling and simulation of the computational model using fuzzy inference systems with MATLAB® Fuzzy Logic Toolbox™.

Chapter 6 deals with the policy decision emergence simulation model PODESIM as a multi-level fuzzy system. PODESIM is a decision diagnostic tool in public policy developed in this research that represents a synthesis of theories and techniques reviewed and that creates a bridge between them.

For this fuzzy system, we develop adequate membership functions and consistent inference rules to proceed with the simulation.

In this chapter, we also carry out a case study to validate the model. Validation is a mandatory step in any modeling and simulation process, and it is carried out using real cases for which we have sufficient and appropriate data to do the simulations. This validation determines whether the model represents the real system and ensures that its assumptions and rules are consistent with the experimental results.

The case study is related to decision emergence during the Cuban Missile Crisis (1962).

The results given by simulation are then analyzed in Chap. 7 and compared with the evidence of the case studied. This chapter also discusses the evaluation of the model and its strengths and limitations.

Finally, we highlight the original contributions and the innovation brought by this research in Chap. 8, and we end up with the conclusion that describes the main steps of this research project.

This conclusion also suggests additional research avenues to improve the PODESIM model and to reinforce modeling and simulation in the field of public policy and public administration.

The ultimate objective is to contribute to the advancement of knowledge and methodologies in the field of computational public policy[1] and to support and strengthen the digital transformation of public policies and governance.

References

De León, P. & Vogenbeck D.M. (2007). The policy sciences at the crossroads. In F. Fisher, G.J. Miller and M.S. Sidney (Eds.), *Handbook of public policy analysis: Theory, politics, and methods* (pp 3–14). Boca Raton, FL: CRC Press, Taylor & Francis Group.

Harrison, N. E. (Dir.) (2006). *Complexity in world politics, concepts and methods of a new paradigm.* Albany. State University of New York Press.

Jervis, R. (1997). *System effects. Complexity in political and social life.* New Jersey: Princeton University Press.

Lempert, R. (2002). Agent-based modeling as organizational and public policy simulators. *Proceedings of the National Academy of Sciences, 99*(3), p. 7195–7196.

Mintz, A., & DeRouen, K. R. (2010). *Understanding foreign policy decision making.* New York, NY: Cambridge University Press.

Mitchell, M. (2009). *Complexity: A guided tour.* New York: Oxford University Press, Oxford.

Morçöl, G. (2003). Complexity and public administration. In *Public Administration Theory Network Annual Conference.* Anchorage, Alaska.

Morçöl, G. (2012). *A complexity theory for public policy.* New York. Routledge.

Moss, S., & Edmonds, B. (2005). Towards good social science. *Journal of Artificial Societies and Social Simulation,8*(4), 13. Available at http://jasss.soc.surrey.ac.uk/8/4/13.html.

Yetiv, S. A. (2011). *Explaining foreign policy: U.S. decision-making in the Gulf Wars* (2nd ed.). Johns Hopkins University Press.

[1]This term is proposed by Ashu M. G. Solo as a new field of research aimed at bringing together computer and mathematical techniques to solve public policy problems. This field includes the principles and methods of the public policy and decision-making field as well as modeling, simulation, optimization, and forecasting. See: Solo, A. M. G. (2014). «The New Interdisciplinary Fields of Public Policy Engineering and Computational Public Policy». In A. M. G. Solo (Ed.), *Political Campaigning in the Information Age.* Hershey, PA: IGI Global.

Chapter 2
Decision Process and Analytical Frameworks—Levels of Analysis and Paradigmatic Evolution

> *[A] "decision" is only a part of a decisional process that began long before the specific decision was made... The momentary act of decision, on which so much of the literature of "decision making" focuses, may be little more than pro forma.*
> *Green, 1966, p. 205*

Analysis of policy decision processes is necessary to explore and explain the reasons and the circumstances that lead to decision-making. According to Hassenteufel (2011), decision-making analysis is both the most obvious and the most problematic aspect of public policy analysis.

Decision-making is such an important aspect of public policy that it raised several epistemological issues related to the understanding of the decision process and the decision-making circumstances. These epistemological thoughts evolved conceptually and methodologically to consider an increasing number of various variables. The diversity of the variables that underly the decision-making makes the decision process more complex, but it also contributes to enrich decision analysis approaches and to grasp a detailed picture of the decision process and the reasons behind the choices made by decision-makers.

Understanding the decision process is a very important aspect to interpret and explain policy choices. According to Mintz and De Rouen (2010), the way decisions are made can shape the eventual choice. That is, an actor could arrive at different outcomes depending on the decision process (p. 4). Therefore, analyzing the decision process is a way to understand policy choices and to explain what happens inside the decision-making "black box".

Furthermore, decision analysis is at the core of the field of policy analysis because public policies are the result of a decision and of a decision process that leads to decision-making.

Regarding the paradigmatic evolution, academic works trace back decision analysis in public policies to the rationality concept that considers the decision-maker as a rational actor in the decision process. The rational actor theoretical view was later challenged by the notion of limited rationality (Simon, 1955) and by the concept of incrementalism (Lindblom, 1959).

A. Guidara, *Policy Decision Modeling with Fuzzy Logic*, Studies in Fuzziness and Soft Computing 405, https://doi.org/10.1007/978-3-030-62628-0_2

As further development, other notions related to organizational choice and bureaucratic politics were added to decision analysis and questioned further the rationality paradigm. Especially since these notions involve not only political decision-makers but also a multitude of administrative actors and networks that take part in the decision process (Allison, 1971).

Finally, Cohen, March, and Olsen (1972) introduced cognitive aspects and stochastic dynamics that contribute to the decision processes. These phenomena also constitute an additional contribution that challenges, even more, the rationality paradigm.

The theoretical contributions in the field of policy decision-making represent an advancement of the analytical approaches and paradigmatic progression of decision analysis. It considers several aspects from different perspectives to better capture the diversity and plurality of variables and dynamics of the decision process that explain policy decision-making.

These contributions also shaped decision analysis frameworks or models as a result of this methodological and epistemological evolution. Allison (1971) calls these frameworks and approaches conceptual lenses necessary to describe and explain reality.

2.1 Decision Analysis Frameworks

In this section, we describe the main decision analysis frameworks in public policy. These approaches or models reflect the chronological stages of the paradigmatic and epistemological development of decision analysis in public policy. They address policy decisions in different ways and raise many characteristics of the decision process at different levels of analysis.

We also describe the level of analysis of the decision process considered by each model and establish a gradual progression of the analysis steps.

2.1.1 Rational Decision-Making Model

The rational decision-making model is a classical approach in public policy analysis. It is inspired by "instrumental" rationality, which assumes that a rational actor will make decisions and choices that maximize benefits and minimize any cost. Rational decision-making is based on the following assumptions (Allison, 1971; Allison & Zelikow, 1999; Green & Shapiro, 1996; March, 1994):

- The decision-maker or the actor identified the problem and defined clearly the goals and objectives.
- The actor has complete and objective information and the ability to prioritize and evaluate each alternative to optimize the choice.

- The actor selects the most advantageous choice that maximizes the benefits while minimizing costs, based on measurable criteria.

This analytical framework has been criticized by many due to its inherent weaknesses.

First, the rationality of the actors is challenged because it is hard to define and it is related to individual considerations. The rationality of policymakers may also be limited by institutional rules or by practices and values that reduce their rationality and limit decision-making choices (Allison, 1971; Green & Shapiro, 1996; March & Simon, 1993). Moreover, in the decision process, an actor cannot ignore the political environment characterized by uncertainty, instability, and the lack of full and reliable information. This does not allow for an objective analysis of the problem and the alternatives.

Allison and Halperin (1972) consider that the rational model is incomplete and they question the rationality of decision-makers either. They point out the bureaucratic organizations and claim that the rational model [...] obscures the penitently neglected fact of bureaucracy: the "maker" of government policy is not one calculating decision-maker, but rather a conglomerate of large organizations and political actor who differ substantially about what their government should do and who compete in attempting to affect both governmental decisions and the actions of their government (p. 42).

Moreover, all the assumptions give the model a linear and deterministic orientation that also obscures many influential and decisive factors and variables of the decision process. Among the factors, some refer to individual perception of decision-makers, uncertainty, and risk evaluation that do not lead to an objective assessment of issues, alternatives, and consequences (March & Simon, 1993). Others describe the rational model as an ideal type far from reality and they critic its methodological utility, given the intangible nature of its assumptions. They argue that the rational model makes unrealistic assumptions.

Green and Shapiro (1996), for example, raised the empirical weakness and the falsifiability of the rational model. They argue that the model suffers from methodological pathologies like a post hoc theory development. They also claim that the rational model has not yielded empirically useful results because its applications are theory-driven rather than problem-driven.

They also point out other weaknesses like the abstract formulation of the model's assumptions, and the unmeasurable nature of the entities to which the model refers. In other words, they raise the impossibility to operationalize any variables from the assumptions of the rational model.

Other works raise the influence of individual perceptions and cognitions on decision-making in the context of presumed rationality. Legrand (2004) argues that one of the basic conditions for the decision-maker to be able to make a rational decision in the full sense of the term would be [therefore] to have an objective perception of the operating environment and an integral power of anticipation in his or her game (p. 97). March (1994) also states that although decision-makers try to be rational, they are constrained by limited cognitive capabilities and incomplete information […] (p. 9).

The limitations of the rational actor model raised by the critics prove that this model is not appropriate for a full analysis of the decision process. Therefore, it cannot grasp the levers of policy decision emergence at the sub systemic level of the decision process.

Since the rational model was criticized for its shortcomings since its formulation, the concept of bounded rationality was proposed by Simon (1957) as an alternative to the rational model. The bounded rationality approach considers that the information available to decision-makers is always imperfect to ensure rational behavior, especially since decision-makers are also influenced by cognitive limitations and biases. But this alternative is not useful for this research, since it does not inform us about the sub systemic level of the decision process.

In conclusion, rationality as well as bounded rationality approaches analyze the decision-making at the systemic high-level of the decision process and the final stage of decision-making. At this stage of the process, the problem, the alternatives, and the decision circumstances are already set.

2.1.2 Incremental Model

In parallel with Simon's bounded rationality concept, Lindblom (1959) proposed the incremental model as an alternative to the rational model. Lindblom considers that the rational model is descriptive and theoretical, and it may be convenient to analyze individual decisions. But he argues that the rational model does not correspond to the reality of public policy decision-making, because rationality is often exposed after the decision is made.

According to Lindblom, a political decision is not based on a single decision-maker but it considers several political imperatives and organizational constraints. Therefore, he proposes an incremental decision-making approach that reflects observable reality rather than systematic analyses advocated by the rational model (Gortner, Mahler, & Nicholson, 1993).

Lindblom describes the observable reality as a gradual progression of multiple incremental steps in the decision process as a result of bargaining between various actors through persuasion, discussion, and negotiation. This process results in a decision-making option that represents a consensus between the parties, which implies that decision-makers are willing to make concessions and compromises and to consider only the alternatives that satisfy the status quo (Lindblom, 1959).

Lindblom also argues that constraints due to time, cost, and cognitive limitations prevent decision-makers to consider all possible alternatives and to assess their consequences. For this reason, the decision process requires small steps of change and adjustments based on the existing situation and the experience of decision-makers, which implies a sustained effort from all parties to persuade and advocate the selected options. Indeed, the incremental approach favors small step actions that can be continually reviewed and evaluated (Geyer & Rihani, 2010).

Although Lindblom considers the incremental model useful for decision analysis, his model is criticized because of its "idealism" as it does not consider the concentration of power in the hands of decision-makers. It does not consider either the inequality between the bargaining parties involved in the decision process. Andrews, for example, states that critics of Lindblom assert that incrementalism is the rational procedure sometimes but not all of the time (Breheny & Hooper, 1985; Smith & May, 1980). There are decisions in which expert contributions to substantive rationality are essential, [...]. In such cases, the goodness of decisions depends equally on their procedural and substantive elements (in Fisher, Miller, & Sydney, 2007, p. 163).

Kingdon (2014) also argues that incrementalism describes parts of the process, particularly the gradual evolution of proposals or policy changes, but does not describe the more discontinuous or sudden agenda change (p. 19). Indeed, the incremental model considers that public policy decisions are just a continuation of previous actions with some incremental changes. In other words, all facets that may influence decision-making are not necessarily considered in the decision process and no alternatives are necessarily given priority (Lindblom, 1959). Furthermore, Etzioni argues that in the incremental model, innovative options are not considered and, all things considered, decisions are somehow made blindly (in Gortner et al., 1993).

Also, in its political extent, the incremental model excludes large segments of the population, which promotes inertia that serves the only interests of influential people by ignoring intangible values (Etzioni, 1967; Gawthorp, 1971; in Gortner et al., 1993).

The review of the incremental model demonstrates that it does not raise any aspect of the sub systemic level of the decision process and it is an actor-centered approach.

2.1.3 Bureaucratic Politics Model

In his popular work on the Cuban missile crisis (1962), Allison (1971) introduces a new decision-making analytical approach that he calls the bureaucratic politics model. This model considers the policy decisions as a result of bargaining among bureaucracies and individuals in government positions. Allison argues that public organizations have their own political and organizational motivations and they play an important role in decision-making through interactions among competing units and actors within the machinery of government.

Allison highlights the rivalries between the different bureaucracies that participate in decision-making or influence the decision process. The essence of the bureaucratic politics model is, therefore, to scrutinize what happens within the government organizations and what is the role played by the actors in the decision process.

According to Allison and Zelikow (1999), the actors involved in the decision process are most often guided by self-interests than by strategic and global benefits. They describe government action and decisions as intranational political resultants;

resultants in the sense that what happens is not chosen as a solution to a problem but rather results from compromise, conflict, and confusion of officials with diverse interests and unequal influence; political in the sense that the activity from which decisions and actions emerge is best characterized as bargaining along regularized channels among individual members of government (p. 294, 295).

This statement constitutes the foundation for the bureaucratic politics model that aims to report the role of bureaucracies and public organizations in the decision process. This role is characterized by the administrative rules, by the bureaucratic power play, and by the bargaining and channels of influence within the government.

This model also integrates administrative routine as an aspect of the decision process dynamics. Allison and Halperin (1972) argue that what emerges is also importantly affected by constraints, in particular by the routines of organizations in supplying information and options, and by shared values within the society and the bureaucracy (p. 51).

The bureaucratic model, therefore, raises questions about the identity of the actors and their real influence on decision-making, but also about the relationship and exchanges that exist between the different actors and bureaucracies at several levels. The importance and influence of each unit depend on its position in the power pyramid, as specified by Allison (1971) who argues that where you stand depends on where you sit (p. 176).

Moreover, bureaucracies do not always impose themselves in the decision process. It is political leaders who call on public organizations to seek assistance for information or expertise. However, the information provided by these bureaucratic units can be biased and obey the interests of those who produce it. These units are likely to influence decisions by providing information that is filtered, selected, and presented as reliable and appropriate to decision-makers.

This situation leads to competition between the different bureaucracies that try to put forward their points of view and favor their choices. Therefore, they consider in the first place their organizational and corporatist concerns before any other consideration.

And yet, competition between bureaucracies could sometimes be beneficial. The involvement of several organizations in the decision-making process can result in multiple advocacies of rival choices, thus improving the chance that all possible options will be considered (George, 1972; in Kegley, Shannon, & Blanton, 2011, p. 65). Eventually, this bureaucratic dynamic could simply generate a decision that is a compromise or a minimal decision, an arbitration, or the option of unilateral victory of one faction over the others (Legrand, 2004, p. 87), thus reducing the benefits of considering various alternatives.

Among the criticisms of the bureaucratic model, some argue that this approach can lead to the overvaluing of the weight of administrative units and their corporate ethos, to the detriment of the weight of the final decision-makers (Legrand, 2004, p. 87). Moreover, because each unit involved seeks to promote its interests, the result is that different groups, each pulling in different directions, produce an outcome (or even a result of conflicting preferences and unequal power) distinct from what any person or group wanted (Allison, 1971, p. 145). Besides, bureaucratic units rely on organizational routines that do not allow all possible options to be explored, as they are likely to operate according to established subsequent approaches. As a result, some factors and variables that may influence the decision process can be simply ignored.

Finally, regarding the empirical potential of the bureaucratic model, Michaud (1996) raises many of the criticisms of this approach, some are even stated by Allison himself. Michaud concludes that if there is one criticism where the theory of bureaucratic politics is more vulnerable, it concerns the methodological deficiencies. The original nature of the model describes a situation that is very difficult to operationalize [...], especially because of the difficult access to the information, necessary for operationalization as well as the distorting imprint that time can induce (p. 779).

By focusing on the role of bureaucracies in the decision process, this model overlooks several factors. It is a prism through which it is possible to analyze the influence of bureaucracies in the decision process. But, alone, this model is not sufficient to explore the decision process and its complexity. In an updated edition, Allison and Zelikow (1999) conclude that the three models discussed in their book must be combined to analyze decision-making, namely the rational model, the organizational model, and the bureaucratic politics model, which confirm the weaknesses of each model.

To this, we add that the bureaucratic model analyses the decision-making at the systemic level and it is also an actor-centered approach. It does not raise the properties of the decision process at a detailed level where there are factors and dynamics at play that dictate sometimes the evolution of the decision process and the policy decision emergence.

2.1.4 Garbage Can Model

To complete this review of the theoretical development of policy decision analysis frameworks, it is essential to mention Cohen, March, and Olsen's work regarding decision-making.

These authors developed an approach called the Garbage Can Model (GCM) to explain the decision-making dynamics within academic organizations portrayed as organized anarchies. Cohen, March, and Olsen (1972) argue that to understand processes within organizations, one can view a choice opportunity as a garbage can into which various kinds of problems and solutions are dumped by participants as they are generated (p. 2). According to them, there are as many problems waiting for solutions as solutions waiting for problems. Since problems are sometimes not entirely determined, stakeholders then propose solutions to potential problems.

This model states that decision-making is a process characterized by ambiguity and uncertainty. Actors do not always have clear and well-defined objectives. They do not necessarily master decision-making dynamics. Therefore, they are not able to consider and evaluate all options and alternatives.

The Garbage Can Model portrays three independent streams that flow disconnected from each other: a stream of problems, a stream of solutions, and a stream of participants such as decision-makers. These three streams only converge when a fourth stream, a stream of choice opportunity, arises. The convergence of the streams may lead to decision-making. Thus, the decision is a random result of the dynamics of the streams that no one controls.

Besides, Cohen, March, and Olsen state that the number of participants and the frequency of their interventions are also fluctuating. This makes it difficult to capture the whole process, especially because the actors' participation is uncertain. Furthermore, the interdependencies between actors and the inequality of their resources are not considered, which leads to more anarchy in the process (Hassenteufel, 2011).

Moreover, according to Gortner et al. (1993), Cohen, March, and Olsen conclude that decision-making is not a process for achieving specific objectives by choosing means that are considered optimal. Therefore, they argue that the Garbage Can Model is a descriptive approach and a model of bureaucratic decision-making that has not been sufficiently studied to explore its analytical potential in the public policy field.

Finally, Hassenteufel (2011) concludes that the Garbage Can model cannot be used as an accurate representation of decision-making processes. However, he pointed out that this model underlines the need to consider the multiplicity of actors involved in the decision-making. Hassenteufel states that the Garbage Can Model also raises aspects that affect the dynamics of decision-making, including the randomness, the plurality of actors, and the complexity of situations that arise at a certain point in the decision process. However, he considers that decision-making still depends on the simultaneous combination of various factors over time and on the sorting that follows this combination (March, 1994). These random characteristics do not allow any operationalization of real cases.

Regarding decision-making and time factor, March (1994) states that in an environment characterized by complex interactions among actors, solutions, problems, and choice opportunities, the simplest source of order is that of time (p. 198). March also argues that in a garbage can process, it is assumed that there are exogenous, time-dependent arrivals of choice opportunities, problems, solutions, and decision makers […] almost any solution can be associated with almost any problem—provided they are evoked at the same time (p. 200). He concludes that the results produced by the system depend on the timing of the various flows and on the structural constraints of the organization (p. 201). However, this does not add any real input to deal with the decision process and with the policy decision emergence.

The model is consequently an aggregate of several decision-making approaches. It refers to the rationality of some actors who explain their actions in terms of alternatives and consequences of their preferences, while others choose satisfactory options, which are actions that are rather characterized by limited rationality.

In addition, March raises the unstable nature of institutional rules and arrangements, which makes some decisions to appear adequate in some circumstances and inadequate in other situations due to an organizational influence in the decision process.

In his work, March also referred to the political decision-making and raised the opportunity of simulation of the Garbage Can Model through the definition of certain variables of the model's streams. Despite this development, the model remains theoretical and difficult to operationalize to deal with real cases. A consistent determination of its variables to proceed with a simulation is impossible.

Cohen, March, and Olsen (1972) conclude that the Garbage Can Model can best be described as a collection of ideas rather than a coherent structure. To this, Daft (2010) adds that the unique characteristics of the garbage can model is that the decision process is not seen as a sequence of steps that begins with a problem and ends with a solution [...] Decisions are the outcome of independent streams of events within the organization (p. 471).

However, March (1994), referring to other perspectives, conclude that the idea of decision making gives meaning to purpose, to self, to the complexities of social life [...] It is hard enough to make sense of the simple things without discovering they are really not as simple as they look (p. 271, 272). In stating this, March raises the complex nature of the decision process in addition to the dynamics already highlighted by the Garbage Can Model, but without investigating this complexity.

Understanding the complexity of reality and the meaning of things that seem simple requires a multidisciplinary approach such as complexity theory. Axelrod et al. (1995) and Gilbert (1995), already advocated the need to carry out decision analyses using new methodological tools, under virtual conditions, to understand the inherent complexity of decision processes.

2.1.5 Chapter's Conclusion

In this chapter, our objective is not to review all decision-making models in public policy, but to discuss the most popular approaches and to assess their relevance concerning the sub systemic environment of the decision process and the decision emergence.

The review of the analytical approaches in public policy sheds light on two fundamental aspects: the first is related to the methodology that each model has developed to understand decision-making, and the second concerns the paradigmatic and epistemological evolution of decision analysis in public policy.

Each of the models provides information on some characteristics of the decision-making, but they do not explore the entire decision process. Moreover, they do not address the sub systemic environment of the decision process. These models were developed from an analytical perspective at the systemic level of the decision process. The limitations of each analytical model led to subsequent developments and this progression constituted the paradigmatic evolution of decision analysis in public policy.

Each approach reviewed in this chapter contributes to the progression of the analysis of the decision-making by adding a complementary layer and details to the decision process. The objective of each contribution is to capture more dynamics of the decision-making process at a different level.

Our review sheds light on the various aspects of the decision process, from the rational decision-making model, which is by nature deterministic and linear, to the Garbage Can Model that raises the inherent complexity of the decision process.

Rationality, the foundation of the rational decision-making model in public policy analysis, corresponds to the instrumental rationality of the positivist paradigm. The limitations of the rational model have led to subsequent approaches. Lindblom's incrementalism (1959) is a response to the increasing dynamics that characterize the decision process in public policy.

Later, Allison (1971) pointed out that in any policy decision, there are processes that compete and interact. One of these processes is represented by bureaucratic politics. Finally, the Garbage Can Model introduced new complementary notions to explain the decision process. This model highlights the complex nature of this process that includes uncertainty, vagueness, randomness, unpredictability, a variety of entities, and finally the absence of any central control over the decision process.

The Garbage Can Model also adds new variables to the decision process, such as problems, solutions, and choice opportunities. However, these variables deal with the systemic or macro-level behavior of the decision process, besides being descriptive and narrative and not operationalizable.

In conclusion, the analytical approaches contributed to the progression of the decision-making analysis and this progression constitutes an accumulation of paradigmatic layers that reveal continuity rather than rupture.

This progression began with the certainty that characterizes the rationality of decision-making. Then, it addressed the compromise between actors in the decision process, and finally, it raises the complexity of the decision process (Geyer & Cairney, 2015). However, this progression does not propose so far methodologies to deal with the complexity of the decision process nor with the sub systemic level of this process where the decision starts to emerge.

Figure 2.1 illustrates the paradigmatic progression of decision analysis since the 1950s.

The paradigmatic evolution of decision analysis shows that the classical analytical approaches only analyze the systemic or the macro-level of the decision process. This analysis only describes the global behavior of the process, that of the actors and decision-makers. These analytical approaches, in a certain way, narrow the decision process to its final stage when the decision is being made.

Simon (1977) already stated that this way of deterministic and linear analysis could not be supported. He argues that all the images falsify decision by focusing on the final moment (p. 40). Besides, it is established that the decision is the result of a process that represents what happens in decision-making prior to choice (Yetiv, 2011, p. 202).

What happens before the final moment is essentially determined by the dynamics of the sub systemic environment of the decision process. None of the analytical approaches deals with this pre-decisional environment that generates the decision circumstances, called the policy decision emergence in this research. To summarize

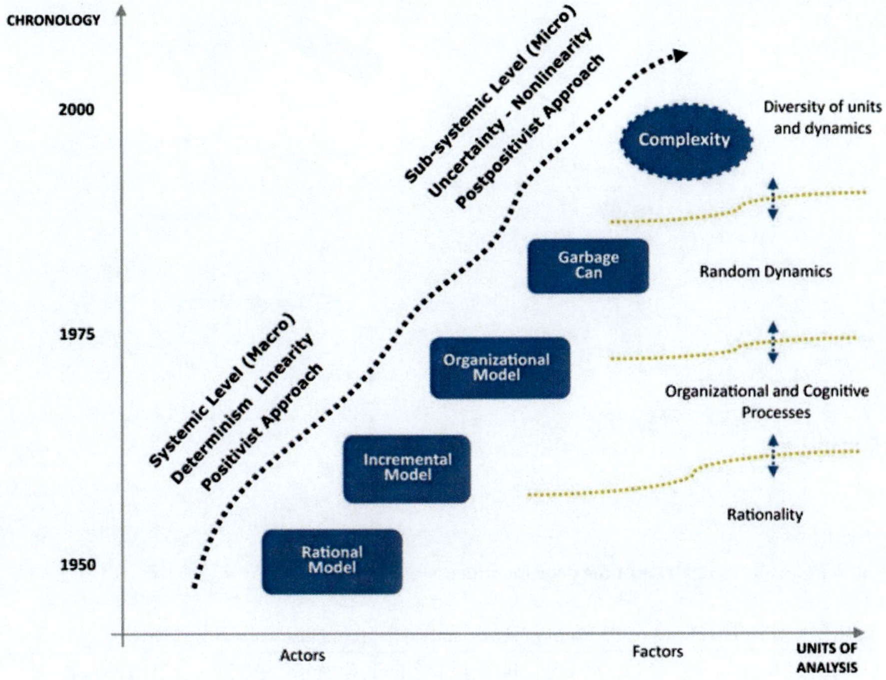

Fig. 2.1 Paradigmatic evolution of decision analysis approaches

this difference, we can argue that the systemic analysis of the decision-making aims to explain how the system behaves at the macro-level. Whereas the sub systemic investigation aims to identify the factors and dynamics that underly the system's behavior. In other words, it aims to explain why the system behaves the way it behaves.

Figure 2.2 presents a synthesis of the concepts developed in this chapter regarding the levels of analysis of the decision process. It shows the exploration of the decision process by classical actor-centered analytical approaches and the progression toward the complex sub systemic environment of this process that considers a variety of factors and dynamics. This progression takes place from the systemic or macro-level to the sub systemic or micro-level.

Table 2.1 summarizes our conclusions regarding the characteristics of the various levels of analysis of the decision process.

This project aims to explore the sub systemic or micro-level of the decision process. This environment is a complex system made of various entities and it is the source of the decision emergence in public policy.

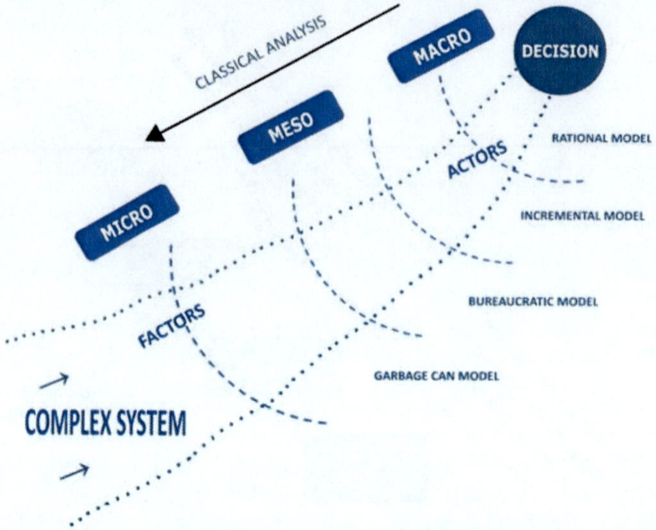

Fig. 2.2 Levels of analysis of the decision process

Table 2.1 Characteristics of decision process analysis approaches

Level	Approach	Level of analysis, nature, and number of entities	Dynamics-control	Nature of information	Type of reasoning
MACRO	Rational	SYSTEMIC Actors (Decision-makers) Limited number	Very weak (centralized control)	Complete Clear Specific	Linear Deterministic Direct
	Incremental	SYSTEMIC Actors (Decision-makers) Limited number	Weak (centralized control)	Complete Clear To be specified	Linear Deterministic gradual
MESO	Organizational	SYSTEMIC Actors (Decision-makers) Bureaucrats Medium number	Medium (fragmented control)	Incomplete Inaccurate	Gradual Uncertain
	Garbage Can	SYSTEMIC Actors Dynamics Variables High number	Strong (fragmented control)	Incomplete Inaccurate uncertain	Gradual Non-linear
MICRO	Complex system	SUB SYSTEMIC Variables Factors Dynamics Very high number	Complex stochastic (No control)	Incomplete Inaccurate uncertain Vague	Approximate Non-linear Uncertain Ambiguous Vague Indeterminate

2.2 Decision Emergence as a Complex System

The review of the analytical models demonstrated that the sub systemic level of the decision process is an environment that is not considered by the analytical approaches despite its importance for the policy decision emergence. This sub systemic level is made of a variety of variables and factors and it is still unexplored.

Few analytical approaches raise the complex nature of the sub systemic level of the decision process but they do not provide methodologies to identify its characteristics and dynamics. In conclusion, the public policy field does not offer any tool to deal with the sub systemic level of the decision process. This environment is a complex system that has properties that correspond to the policy decision emergence concept. This emergence starts at the micro-level of the policy decision process and this sub systemic environment is made of multiple components that interact in a non-linear and non-deterministic way. These dynamics generate the global behavior of the system that cannot be predicted or deduced from the individual properties of the components.

Besides, a complex system is also characterized by the absence of direct causal links, of central control, and of permanent equilibrium state that makes it impossible to predict its evolution. The decision emergence that takes place at the sub systemic environment of the decision process has all the characteristics of a complex system.

We, therefore, need to explore the complex systems theory to identify the tools provided by this field to characterize and operationalize decision emergence.

References

Allison, G. T. (1971). *Essence of decision: Explaining the Cuban missile crisis*. Boston: Little, Brown

Allison, G. T., & Halperin, M. H. (1972). Bureaucratic politics: A paradigm and some policy implications. *World Politics, 24,* 40–79.

Allison, G. T., & Zelikow, P. (1999). *Essence of decision: Explaining the Cuban Missile Crisis* (2nd Ed.). Longman: Addison-Wesley Educational Publishers Inc.

Axelrod, R., Mitchell, W., Thomas, R. E., Bennett, D. S., & Bruderer, E. (1995). Coalition formation in standard-setting alliances. *Management Science, 41*(9), 1493–1508.

Cohen, M. D., March J. G., & Olsen J. P. (1972). A Garbage can model of organizational choice. *Administrative Science Quarterly, 17*(1), 1–25.

Daft, L. R. (2010). *Organization theory and design* (10th ed.). Mason, OH: South-Western Cengage Learning.

Etzioni, A. (1967). Mixed-scanning: A "third" approach to decision-making. *Public administration review*, 385–392.

Geyer, R., Cairney, P. (Dir.) (2015). *Handbook on complexity and public policy*. Edward Elgar Publishing Limited.

Geyer, R., & Rihani, S. (2010). *Complexity and public policy: A new approach to twenty-first century politics, policy and society*. Abingdon: Routledge.

Gilbert, N. (1995). Emergence in social simulation. In N. Gilbert, R. Conte (Ed.), *Artificial societies: the computer simulation of social life* (pp. 144–156). London: UCL Press.

Gortner, H. F., Mahler, J., & Nicholson, J. B. (1993). *La gestion des organizations publiques*. Presses de l'Université du Québec.
Green, D. P., & Shapiro. I. (1996). *Pathologies of rational choice theory. A critique of applications in political science*. Yale University Press.
Hassenteufel, P. (2011). *Sociologie politique, l'action publique*. Paris: Armand Colin.
Kingdon, J. (2014). *Agendas, alternatives and public policies* (2nd ed.). Essex: Pearson Education Limited.
Legrand, V. (2004). La prise de décision en politique étrangère. Dans C. Roosens, V. Rosoux et T. Wilde D'Estmael (Dir.), *La Politique étrangère: Le modèle classique à l'épreuve* (p. 79–106). Bruxelles: Presses Interuniversitaires Européennes.
Lindblom, C. E. (1959). The science of muddling through. *Public Administration Review, 19*(2), 79–88.
March, J. G. (1994). *Primer on decision making: How decisions happen*. Free Press.
March, J. G., & Simon, H. A. (1993). *Organizations* (2nd ed.). Cambridge, Massachusetts: Blackwell Publishers.
Michaud, N. (1996). Graham Allison et le paradigme bureaucratique: vingt-cinq ans plus tard est-il encore utile ? *Études internationales, 27*(4), 769–794.
Mintz, A., & DeRouen, K. R. (2010). *Understanding foreign policy decision making*. New York, NY: Cambridge University Press.
Simon, H. A. (1955). Behavioral model for rational choice. *Quarterly Journal of Economics, 68*, 89–118.
Simon, H. A. (1957). Models of man, social and rational: Mathematical essays on rational human behavior in a social setting. New York: John Wiley and Sons.
Simon, H. A. (1977). *The new science of management decision* (3rd Ed. rev, 1st Ed, 1960). Prentice-Hall, Englewood Cliffs, NJ.
Yetiv, S. A. (2011). *Explaining foreign policy: U.S. decision-making in the Gulf Wars* (2nd Ed.). Johns Hopkins University Press.

Chapter 3
Complex Systems and Public Policy

So far as the laws of mathematics refer to reality, they are not certain. And so far as they are certain, they do not refer to reality.
Albert Einstein

Complex systems theory, or complexity theory,[1] is generally presented as a new approach to science in which we identify (and then explain) systems or processes that lack the order and stability required to produce universal rules about behavior and outcomes (Cairney, 2012a, 2012b).

Complexity theory is considered as a revolutionary turning point in science in comparison with traditional scientific approaches. Some describe complexity theory as a challenge to positivist approaches in political science, and to the "paradigm of order" which is characterized by determinism, reductionism, linearity, and prediction (Geyer & Rihani, 2010). Geyer and Cairney (2015) argue that complexity theory represents a major problem for positivist studies driven by a belief in objectivity and linear cause and effect (p. 7). They argue that complexity theory is a new scientific paradigm and a new way of thinking, understanding, and studying the world in perpetual fluctuation. They also claim that complexity theory suggests that what we call "real" is a brief snapshot of a world that is always in flux. Consequently, to advance our study of policymaking in a complex world, we need to understand the problematic ways in which policymakers see and respond to it (p. 6).

Cairney (2012a) states that complexity theory represents a profoundly new way to examine politics; a paradigm shift in the social sciences that will help replace rational choice theory and shift our focus of explanation from individualistic to holistic accounts (p. 355). Furthermore, Collard, Verel, and Clergue (2013) argue that complexity theory requires a renunciation of preconceived ideas and a confrontation with new challenges [...]. It is necessary to abandon the cause-and-effect relationship between a local plurality and a global emergence (p. 19).

Morin (2005), on his side, highlights the complexity paradigm shift and its consequences on science and knowledge in general. He states that what affects a paradigm,

[1] Most works on complexity state that there is no unified theory of complexity. It is rather the concept of complexity and the properties of systems identified as complex that induce methodologies and techniques to understand these systems and their dynamics, which could vary from a field to another.

A. Guidara, *Policy Decision Modeling with Fuzzy Logic*, Studies in Fuzziness and Soft Computing 405, https://doi.org/10.1007/978-3-030-62628-0_3

i.e. the keystone of an entire system of thought, affects ontology, methodology, epistemology, logic, and consequently practice, society, and politics (p. 73).

From a methodological point of view, complexity theory is an interdisciplinary field for understanding complex systems and behaviors using specific methods and techniques for this purpose. The study of complex systems has achieved great progress both in terms of methods and formalization. The scope of complex systems theory is very broad and covers all areas, including policy analysis (Bourcier, Boulet, and Mazzega, 2012). Indeed, from the mid-1990s, the inherent complexity of public policy and decision processes started to be a policy research topic. Geyer and Rihani (2010) argue that what has been emerging since the 1990s has been a blossoming of complexity-based works in a variety of policy sub-areas (p. 31).

That being said, the complex nature of political processes is developed within the field of public policy, through approaches that have shaken up the dominant definition of public policy, namely that public policy is what governments choose to do or not to do.

Morçöl (2010) argues that more sophisticated conceptualizations of public policy processes reject this simplistic and instrumentalist view and acknowledge that policy processes are complex. He cites as examples the work carried out by Ostrom on the Institutional Analysis and Development framework (IAD), by Sabatier and Jenkins-Smith concerning Advocacy Coalition Framework, and by Koppenjan and Klijn on Network Governance Theories. Morçöl (2010) argues that these three examples among others conceive public policies as multi-layered systems populated by individuals and aggregates of actors, and in which public policy processes are designed in terms of relationships between the micro-level and macro-level. He concludes that the recognition of the complexity of policy processes by these theorists opens up the space where a complexity theory of public policy can be built (p. 52).

The works mentioned by Morçöl facilitated the recent introduction of complexity theory into the public policy and decision analysis field. Geyer and Rihani (2010), for example, claim that studying complexity in various public policy areas shows how quickly complexity has moved from the fringes of social science theory to the day-to-day of policy decision-making (p. 31).

More explicitly, Colander and Kupers (2014) put forward an important argument regarding policy issues and complexity. They claim that the interconnected nature of the problems we are facing doesn't fit the standard frame's simplified assumptions. Complexity science came about in an attempt to understand these kinds of highly interconnected systems. The complexity frame provides a new way to look at problems, and it is already starting to influence policy discussion (p. 17). However, they wonder how to decide whether a method from complexity theory is more appropriate than a conventional approach. They propose that the more dynamically and tightly interrelated the parts are, the more likely the complexity frame will be the more useful one [...]. They argue that in complex system, in principle, everything influences everything else (p. 15).

To clarify the fundamental difference between the classical analytical method and the complex approach, Colander and Kupers (2014) also specify that in the standard social science policy model, the dynamic interconnections among agents in the society are suppressed and their importance hidden by the assumptions of the model. In the complexity frame, they are not (p. 15). However, they warn about the fact that models of complex systems represent limits that one must tolerate since it is technically impossible to grasp and represent all the dynamics and interactions of a complex system.

Furthermore, they add that in the complexity frame, scientific models provide vision for policy, not an answer for policy (p. 16). They insist on the fact that including the complexity frame is not just a nice addition to the policy menu; it's an absolute necessity (p. 17). They also specify that the fact that order can emerge from the bottom up has enormous implications for policy that have not been built into the standard policy narratives (p. 105).

Lastly, they argue that this phenomenon of sudden emergent phase transition is an important pattern of complex systems that undermines linear empirical estimating processes, such as linear regression analysis (p. 121). Therefore, they regret that all too often policy is conducted on the basis of the assumption that past trends are an indication for the future, and the relationships do not exhibit such phase transitions (p. 121).

Finally, according to Colander and Kupers (2014), the complexity frame offers new ways for policymakers to search for pragmatic answers to our intractable problems (p. 17), which bridge complexity theoretical development to the public policy practice.

Other works also argue that public policies are becoming a subject for complex systems theory insofar, as traditional "linear" policies fail to address the complexity of the networks of actors and norms involved in the legislator's choice.[2] These various works demonstrate the complexity of contemporary public policies and suggest using complexity theory to address policy issues and decision processes as complex systems.

Additionally, Rhodes and Ross (2010) argue that complexity is a source of innovation because it allows things that would not otherwise be combined to be brought together in unexpected ways (p. 192). Based on several studies, these authors claim that the introduction of complexity theory techniques [in the field of public policy] can reinvigorate policy analysis. As for Bourcier, Boulet, and Mazzega (2012) who encourage larger use of complexity theory and its methodologies in the field of public policy. They assert that the analysis of public policies, considering all frameworks, represents certain limits of description and explanation, and shows that the field of public policies can be described as complex. In return, the methodological tools dedicated to the study of complex systems have therefore a vast field of investigation (and application) still very little explored (p. 127).

Regarding recent advancements of interest for this research, the field of complex systems has made great progress with the development of new tools and methods. These methods deal with the complexity of political processes and they overcome the

[2]See: http://iae.univ-lyon3.fr/2e-atelier-de-recherche-complexite-et-politiques-publiques-917066.kjsp.

linear and reductionist approaches, thanks to advances in computational methodologies. Colander and Kupers (2014) argue that the complexity frame is developing now because of advances in computational and analytic technology, which have made it possible to formally analyze issues that previously were too complicated to analyze. Advances in the complexity toolkit are allowing scientists to formally conceptualize relationships and processes that previously couldn't be formalized, or seemed so blurry as to be seen as beyond science (p. 48). In other words, complexity theory makes possible the exploration of the processes that could not be addressed with conventional tools. It is a powerful theory that offers a comprehensive understanding of political processes by demystifying the complexity of the real world.

3.1 Properties of Complex Systems

The concept of complexity in humanities and social sciences is not new. Simon (1962) describes a complex system as [...] one made up of a large number of parts that interact in a non-simple way. In such systems, the whole is more than the sum of the parts [...] given the properties of the parts and the laws of their interaction, it is not a trivial matter to infer the properties of the whole (p. 468).

Mitleton-Kelly (2003) also states that the theories of complexity provide a conceptual framework, a way of thinking, and a way of seeing the world (p. 3). Besides, Sawyer (2005) adds that complexity theory refers to both systems theory and emergence theory. He defines the emergence as the processes whereby the global behavior of a system results from the actions and interactions of agents (p. 2).

The fundamental idea underlying any complex system is indeed the emergence of a global phenomenon, at the macro-level of the system, generated by the dynamics among its components (Holland, 1995). A complex system is therefore an intertwined set of entities that is difficult to break down into separate parts or independent processes, because of the relationships and the interactions among its various components (Gershenson and Heylighen, 2005). The dynamics of such a system determine its evolution and its global behavior.

Others claim that complex systems display properties that cannot be understood by just looking at the properties of the individual components, but are created as a result of the structure and organized interactions between these components (Nikolic and Kasmire, in van Dam, Nikolic, & Lukszo, 2013, p. 18).

Moreover, Collard, Verel, and Clergue (2013) claim that when a system can no longer be totally "divided" into independent parts, it expresses a transition from complicated to complex: the resulting global behaviors become emergent, and the interdependence of components is then expressed in a network of non-linear interactions that is the source of complexity (p. 3). They also point out that the existence of many entities in a system is not a sufficient condition for the emergence of complexity (p. 3). In other words, there must be a dynamic within the system that generates the behavior at the macro-level as a consequence of that dynamic. This is called the emergence phenomenon. Weisbuch and Zwirn (2010) argue that we speak of emerging behavior in the sense that, at the global level, unexpected or counter-intuitive properties appear.

Some have described the emergence as a "generative" behavior, i.e. emergence of global behaviors generated through a bottom-up process that cannot be predicted from the individual properties of the component (Weisbuch & Zwirn, 2010; Epstein, 2006). That means that the same configuration at the micro-level does not necessarily generate the same global behavior at the macro-level because the dynamics among the components of the system can be different.

Such properties of complex systems require different methods and tools than linear methods or conventional analytical frameworks. According to Weisbuch and Zwirn (2010), the evolution of complex systems does not follow physical or mathematical laws that are easy to determine. The behavior of complex systems may be subject to phase transitions as a sudden large change, or a qualitative transformation due to a small variation in one of the control parameters (p. 2).

From another perspective, complex systems are typically far from equilibrium even if they tend towards precarious forms of equilibrium. According to Colander and Kupers (2014), in complex systems, nonlinear systems can have many different possible outcomes, not just a single one (p. 53). The absence of a permanent equilibrium state is another important characteristic of complex systems. Colander and Kupers (2014) claim that complex systems have a tendency to change suddenly, as the system shifts from one equilibrium to another (p. 53). They also specify that at best, complex systems can be influenced—not controlled (p. 104).

Another characteristic of complex systems is that they are open environments without central control of the components. Therefore, they can integrate new entities (Le Moigne, 1999) and ignore components depending on the context and the needs of the system.

Beyond these necessary explanatory notions for the understanding of complex systems, our objective is to find how this theory can help to deal with the policy decision emergence, since it is a complex system, as described in Section 2.2.

We demonstrated that complex systems cannot be handled with traditional or classical analytical models and that they require appropriate approaches that must encourage innovation. Koliba and Zia (2012) also argue that traditional models are not able to address all aspects of complex policy interactions, which indicates the need for the development of hybrid simulation models consisting of a combinatory set of models built on different modeling theories (in Janssen, Wimmer, and Deljoo, 2015, p. 7). Furthermore, Siegfried (2014) adds that the analysis of many systems, processes and phenomena is often only feasible by developing simulation models and executing them using vast amounts of computer power. He concludes that due to the complexity to be represented within models and increasingly detailed representation of dynamic behavior, simulation is often the only choice for analyzing such models (p. 1).

As for Janssen et al. (2015), who argue that in policy implementation and execution, many actors are involved and there are a huge number of factors influencing the outcomes; this complicates the prediction of the policy outcomes. Simulation models are capable of capturing the interdependencies between the many factors and can include stochastic elements to deal with the variations and uncertainties (p. 6, 7). They also state that simulation models do not rely on mathematical abstraction and are therefore suitable for modeling complex systems (p. 7). These remarks are supported by Le Moigne (1999) who argues that to understand and therefore give meaning to a complex system, we must model it to build its intelligibility (understanding) (p. 19).

With all these arguments, we conclude that modeling and simulation is an appropriate method to deal with policy decision emergence as a complex system.

3.2 Modeling and Simulation of Complex Systems

Modeling and simulation[3] constitute an alternative methodology to traditional deterministic approaches that are inadequate for understanding complex systems. Siegfried (2014) argues that the basic terms (complex) system, modeling and simulation are often used together and indeed, they are deeply connected (p. 11). Moreover, many argue that in the social sciences, generally speaking, experimentation is impossible and observation is uncertain, and we must use modeling to represent reality from scattered elements that we are able to collect. Also, Le Moigne (1999) states that modeling a complex system is first and foremost modeling a system of actions (p. 45). As for Quéau (1986), who argues that between theory and experience, between mathematical formalization and phenomenological observation, simulation offers a third way: algorithmic exploration (p. 147).

In a similar perspective to that adopted in this research, Edmonds and Gershenson raise the interest of modeling and simulation by stating that for policy and decision-making, models can be an essential component, as models allow the description of a situation, the exploration of future scenarios, the valuation of different outcomes and the establishment of possible explanations for what is observed (in Geyer & Cairney, 2015, p. 205).

These remarks support the choice of modeling and simulation as the appropriate methodology for dealing with the policy decision emergence as a complex system. Axelrod (1997) points out that simulation makes possible the understanding and identification of policy levers, which is at the heart of this research project. Let us review in detail the two aspects of this methodology.

3.2.1 *Modeling*

According to Sokolowski and Banks (2009), modeling is the act of building models that represent approximations of the real world. Sterman (2000) clarifies the aim

[3]For a historical and paradigmatic review of modelling and simulation, see: John A. Sokolowski & Catherine M. Banks (2009). Modeling and Simulation for Analyzing Global Events. Chapter 1.

of modeling by stating that models are developed to facilitate the solving of real-world problems. He adds that modeling is also a part of the learning process. It is an iterative, continual process of formulating hypotheses, testing and revision, of both formal and mental models (p. 83).

While claiming that there are many definitions and interpretations of the term "model", depending on the discipline and the epistemological point of view, Treuil, Drogoul, and Zucker (2008) argue that a model is an abstract structure that allows us to understand how a reference system works by answering a question about it. As a simple representation of the system, a model is based on general theory and is expressed in a specific language called modeling language (p. 1). Any model is therefore an abstract simplification of the system it represents. It does not include all the properties of the real system but allows what is complex to be represented in an understandable way. Morçöl (2012) states that because it is not possible to describe all the details of complex systems, researchers use simplified models that capture their essential elements (p. 234).

This compromise in the representation of reality means that a model requires the formulation of hypotheses to adapt and model the reality. Railsback and Grimm (2012) claim that we build and use models to solve problems or answer questions about a system or a class of systems (p. 4). This gives a practical aspect to the concept of model. However, these authors specify that any model must pursue a specific objective. They argue that the objective of a model is decisive. Moreover, the objective guides the choice of entities that constitute the model and the degree of detail of the modeled system, especially when it comes to public policies and decision-making.

Furthermore, Edmonds and Gershenson (2013) argue that models can be an essential component, as models allow the description of a situation, the exploration of future scenarios, the valuation of different outcomes, and the establishment of possible explanations for what is observed. Therefore, modeling requires theoretical knowledge and savoir-faire related to the field of study to determine the components of the model that represents the system, and to define its structure and its dynamics. It also requires an appropriate computing application to proceed with the simulation of the model and to generate results.

For that, the designer of a model must combine rigor and judgment. Rigor in the choice of model's components, structure, and the level of details, and judgment in the formulation of assumptions and rules that govern the model's behavior. According to Hoffman, the more detail that is included, the harder it is to follow the dynamics of the model [...]. Simpler, generic models capture fundamental dynamics (in Klotz and Prakash, 2008, p. 198). In other words, the principle of parsimony must be respected to design models with a minimum but sufficient components and assumptions. In the end, modeling should always represent a compromise between the abstract representation of reality, the simulation requirements, and the degree of expected precision.

In this regard, Saunders-Newton states that a good model is not necessarily the one that is an isomorph of the actual system, but is rather one that can be used to perform crucial experiments that are useful in the context of an argument or problem [...] (in Harrison, 2006, p. 171).

3.2.2 Simulation

Treuil et al. (2008) argue that, in general, simulation is the activity from which, according to specific objectives, and with the help of an experimental mechanism (called a simulator), an experimenter disrupts the dynamics of the model to understand its functioning. Simulation is therefore a kind of abstract experimentation (Axelrod, 1997) or a virtual laboratory that reveals the logic underlying the behavior of a complex system and its components. Furthermore, Epstein and Axtell (1996) claim that simulation allows exploring the overall behavior of a system at the macro-level, based on interactions between the components of that system at the micro-level. It has an exploratory and explanatory purpose. Thus, it allows us to test, understand, and explain situations that represent reality.

Moreover, Helbing (2012) argue that if properly done, computer simulations can deliver reliable results beyond the range of analytical tractability (p. 26). Simulation is, therefore, a research and an experimentation technique using a simulation engine, i.e. a computing technique.

However, if Mace and Pétry (2000) argue that simulation has several advantages for the researcher, including the advantage of studying a practical or theoretical problem without reference to empirical data, they point out, though, that this technique involves some risks. Indeed, the quality of simulation results varies depending on the skills and caution of the user, especially during model design, since the development and arrangement of mathematical relations of the model result from the researcher's personal choices.

Nevertheless, the benefits of simulation are significant. Indeed, simulation plays a double role as "experimental" support and as a generator of knowledge and theories. Moss and Edmonds (2005) argue that simulation is a powerful tool that can enrich the social sciences by generating and validating empirical evidence and theoretical knowledge. This knowledge is generated through the combination of inductive and deductive approaches, which characterize the simulation of complex systems.

Regarding simulation in the policy field, Hoffman argues that if one could capture essential elements of an actual social system, it would make testing policy scenarios relatively quick and easy (in Klotz and Prakash, 2008, p. 199). Scenario simulation can be very useful for decision-making and predicting political actions, but also and above all, a simulation model can lead to insights into where there might be policy leverage in the real world (Axelrod, 1997, p. 143). This relates not only to policy-makers but also to the public administration that is responsible for implementing policies.

To synthesize the notions of modeling and simulation, we represent in Fig. 3.1 the modeling and simulation process and cycle.

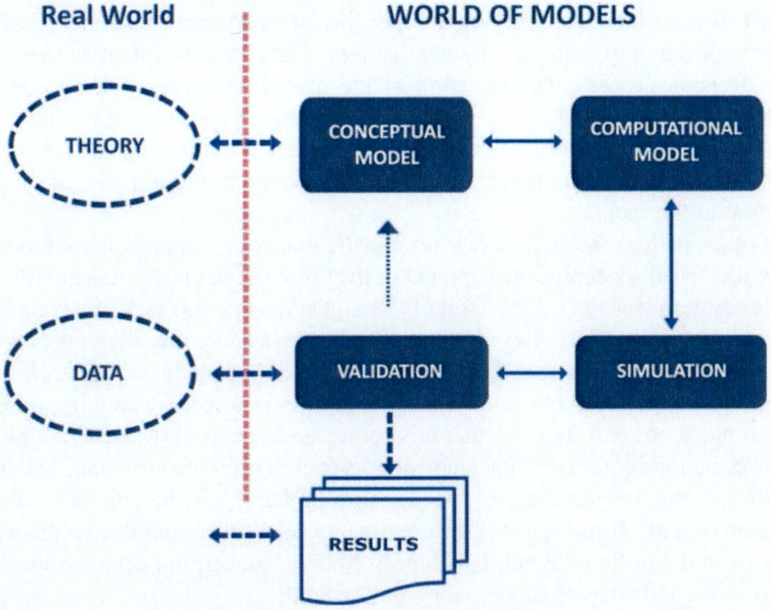

Fig. 3.1 Process and cycle of modeling and simulation

In this process, the modeler is inspired by the theory that describes the reality of a certain system to design a conceptual model that is a representation of that system. The conceptual model must include the main properties of the system under study. These properties include the entities that constitute the system, the rules that govern it, and any other elements considered necessary to represent the real system. The conceptual model becomes a basis for virtual experimentation of the system, i.e. simulation. But to carry out the simulation, the conceptual model must be converted to a computing program that includes the virtual entities of the system, the rules that govern it, and the assumptions related to its evolution, which are developed by the practitioner depending on the related fields. Once the model is built, the simulation under different scenarios reveals the behavior and the evolution of the system and confirm if the model is working as expected. This is called the verification step. Once this step is accomplished, the model is then simulated with data from empirical cases to validate it and to demonstrate that the model can process real cases.

Finally, it is important to keep in mind that all the steps of modeling and simulation take place in a back and forth process between theoretical knowledge, empirical data, and the modeling and simulation cycle. This is to strengthen each step of the cycle, either to improve the model through theoretical concepts, collected data, and results obtained; or to advance theoretical concepts through the results of the simulation.

In this research project, before applying the modeling and simulation methodology to the policy decision emergence, two important preliminary steps need to be completed.

The first step is about identifying the appropriate theory that constitutes a basis and a guide to design the conceptual model that represents the sub systemic environment of the decision process. The structure of the model must be based on consistent theoretical concepts that link the model of this complex system to the public policy field. The second step is to determine the appropriate computing methodology to design the computational model and carry out the simulation for the verification and validation of the model.

However, in this research, there is no specific theory or comprehensive model that deals with the sub systemic environment of the policy decision process qualified as a complex system (see Sect. 2.2). Morçöl (2010) claims that this lack of a comprehensive framework in public policy applications may be because no coherent complexity theory exists in general yet, as Mitchell (2009: 14) acknowledges (p. 52). He argues that the contribution of public policy analysis theories is necessary to address complex political processes and suggests that this contribution be used (Morçöl, 2008).

However, none of the decision analysis approaches presented in Chap. 2 addresses the sub systemic environment of the decision process, nor the complex nature of this environment. To bridge this gap, we use a tool from complexity theory that guides us to the desired layout. It is Stacey Matrix, concerning decision-making in organizations and adapted to the public policy field.

3.3 Stacey Matrix: A Complexity Tool

With a focus to link complexity and management, Stacey[4] developed a matrix to identify the various decision-making situations within organizations. These situations can be simple, complicated, complex, or chaotic. The matrix developed by Stacey is based on two factors that determine the degree of simplicity or complexity in a decision-making situation. These two factors are Certainty and Agreement within the decision-making environment.

Stacey Matrix inspired complexity and decision analysis works in the field of public policy, such as that of Geyer and Rihani (2010) and Geyer and Cairney (2015). These authors argue that the Stacey Matrix is an appropriate tool to explore the nature of the decision process in complex situations.

By relating Stacey Matrix to the public policy field, these authors explain that a high degree of certainty and agreement implies a good knowledge of the public issue that leads to a purely technical decision. In other words, to a rational decision based on reliable data and logical analyses, or even on a well-defined causal relationship among the parameters involved. They cite as an example the Evidence-based Decisions (EBD). These decisions are easily justifiable because they are of a technical nature. They also add that when uncertainty increases and agreement is absent, the decision process tends to become more complex or even anarchic, which

[4]See: Ralph Stacey (2000). Complexity and Management (Complexity and Emergence in Organizations), Routledge.

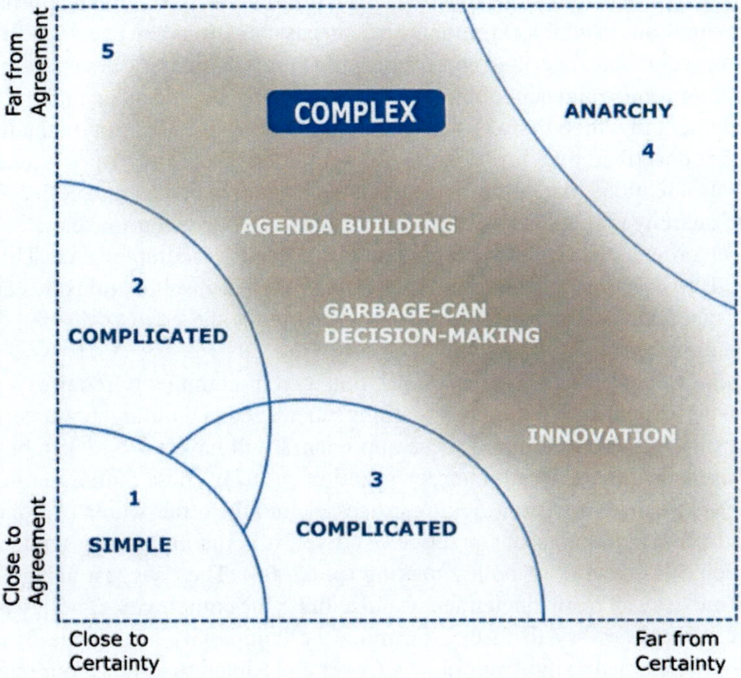

Fig. 3.2 Stacey matrix adapted to policy decision-making (From Geyer & Rihani, 2010; Geyer & Cairney, 2015)

is an extreme situation that leads to denial because it represents an unmanageable situation that decision-makers prefer to avoid.

Figure 3.2 shows the Stacey Matrix and its different decision-making zones that depend on the two factors of Certainty and Agreement.

The different zones in the graphical representation describe different situations of decision-making.

Zone 1 represents a high degree of certainty and agreement. It is an area of order that describes simple situations in which decision-making does not represent real challenges. The decision process is described as technical and rational. In this area, the issues, the solutions, and the outcomes are known. Decisions are easy to make, and solutions easy to implement. Conventional problem-solving techniques and methods, or protocols and procedures, can be easily applied to situations that correspond to this zone.

Zones 2 and 3 represent rather complicated situations.[5] These situations imply various degrees of difficulty depending on the level of certainty about the issue and the agreement among participants in the decision process. However, they remain zones

[5]The term complicated describes a divisible system whose entities have a deterministic behaviour. The overall result of such a system is often predictable.

of order that represent situations that can be analyzed with conventional approaches. These are situations in which extrapolation from past situations can usually be used to predict future outcomes. Furthermore, zone 2 illustrates democratic decision-making based on compromise and negotiation, and zone 3 represents a decision making based on the judgment in which there is agreement on the objectives but not on the means.

Zone 4 is described by a low degree of certainty and a low degree of agreement. It represents situations that are difficult to manage and that can lead to collapse. It is an area of anarchy that leads to denial and avoidance by decision-makers.

Between order and anarchy, there is zone 5, qualified as a complex area. This area corresponds to dynamic and emerging behaviors, where causal relationships are not deterministic. Extrapolation is therefore impossible, as the context of the issue is often uncertain.

According to Geyer and Rihani (2010), this area of complexity requires a range of approaches to deal with complex situations (open agenda building, brainstorming, muddling through, etc.). Some of these approaches will have no evidence base but will be based on expert opinion and/or intuition (p. 67). These authors argue that [...] trying to push all decision-making processes into the orderly area of Zone One is a maladaptive and dangerous practice. And yet, it is the underlying tendency of the traditional framework of policy making (p. 67–68). They suggest using Stacey Matrix to move away from that tendency and address the complexity of policy issues. Therefore, it is necessary to further examine the complexity area (zone 5) of the Stacey Matrix adapted to public policy by Geyer and Rihani to advance our quest for appropriate methodology to deal with the policy decision emergence as a complex system.

In this area, the authors mention two notions that inspire a link between complex decision-making and public policies. These notions are the Agenda building (or Agenda setting), and the Garbage Can Model of decision-making.

In Chapter 2, we demonstrated that the Garbage Can Model of decision-making cannot be operationalized to allow the modeling of decision emergence as a complex system. Moreover, the attempt by its developer to simulate this model is considered as trivial, deterministic, and unrepresentative of the complex institutional processes.[6]

However, the two notions of Agenda setting and the Garbage Can Model are related to the multiple streams theory, developed by John Kingdon. Therefore, the multiple streams theory represents a methodological tool that bridges complexity theory and policy decision emergence. This tool can be used to build a consistent conceptual model.

---

[6]See on this subject: J. Bendor, T.M. Moe et K.W. Shotts (2001). Recycling the Garbage Can: An assessment of the Research Program, *American Political Science Review*, 95, 169–190.

References

Axelrod, R. (1997). *The Complexity of Cooperation: Agent-based Models of Competition and Collaboration*. New Jersey: Princeton University Press.

Bourcier, D., Boulet, R., & Mazzega, P. (2012). *Politiques publiques Systèmes complexes*. Paris: Hermann Éditeurs.

Cairney, P. (2012a). Complexity theory in political science and public policy. *Political Studies Review, 10,* 346–358.

Cairney, P. (2012b). *Understanding Public Policy*. New York: Palgrave Macmillan.

Colander, D. C., & Kupers, R. (2014). *Complexity and the art of Public Policy: Solving Society's Problems from the Bottom up*. Princeton: Princeton University Press.

Collard, P., Verel, P., & Clergue, M. (2013). *Systèmes complexes. Une introduction par la pratique*. Lausanne: Presses Polytechniques et Universitaires Romandes.

Edmonds, B., & Gershenson, C. (2013). Modelling Complexity for Policy: Opportunities and Challenges. Dans *Handbook of complexity and Public Policy*. Edward Elgar Publishing Ltd.

Epstein, J. M. (2006). Generative Social Science: Studies. In *Agent-Based Computational Modeling*. Princeton, NJ: Princeton University Press.

Epstein, J. M., & Axtell, R. L. (1996). *Growing Artificial Societies: Social Science from the Bottom Up*. A Bradford Book.

Gershenson, C., Heylighen, F. (2005). How can we think the complex? In K. Richardson (Ed.), *Managing organizational complexity: Philosophy, theory and application* (Chap. 3, pp. 47–61). Information Age Publishing.

Geyer, R., & Cairney, P. (Dir.) (2015). *Handbook on Complexity and Public Policy*. Edward Elgar Publishing Limited.

Geyer, R., & Rihani, S. (2010). *Complexity and Public Policy: A New Approach to Twenty-first Century Politics Policy and Society*. Abingdon: Routledge.

Harrison, N. E. (Ed.) (2006). *Complexity in world politics: Concepts and methods of a new paradigm*. Albany: State University of New York Press.

Helbing, D. (Ed.) (2012). *Social self-organization: Agent-based simulations and experiments to study emergent social behavior*. Berlin, Heidelberg: Springer- Verlag.

Holland, J. H. (1995). *Hidden Order: How Adaptation Builds Complexity*. New York: Basic Books.

Janssen, M., Wimmer, M. A., & Deljoo, A. (Eds). (2015). *Policy Practice and Digital Science. Integrating Complex Systems, Social Simulation and Public Administration in Policy Research*. Switzerland: Springer International Publishing.

Klotz, A., & Prakash, D. (2008) (Eds.). *Qualitative methods in international relations: A pluralist guide*. UK: Palgrave Macmillan.

Koliba, C., & Zia, A. (2012). *Governance Informatics: Using Computer Simulation Models to Deepen Situational awareness and governance design considerations policy Informatics*. Cambridge: MIT Press.

Le Moigne, J.-L. (1999). *La modélisation des systèmes complexes*. Paris: Dunod.

Mace, G., & Pétry, F. (2000). *Guide d'élaboration d'un projet de recherche en sciences sociales*. Les Presses de l'Université Laval.

Mitchell, M. (2009). *Complexity: A Guided Tour*. New York: Oxford University Press, Oxford.

Mitleton-Kelly, E. (2003). *Complex Systems and Evolutionary Perspectives on Organizations: The Application of Complexity Theory to Organizations*. Bingley, UK: Emerald Group Publishing Limited.

Morçöl, G. (2008). A Complexity Theory for Policy Analysis: An Outline and Proposals. In D. L. F. Dennard, K. A. Richardson, & G. Morçöl (Eds.), *Complexity and Policy Analysis: Tools and Concepts for Designing Robust Policies in a Complex World*. Goodyear, AZ: ISCE Publishing.

Morçöl, G. (2010). Issues in reconceptualizing public policy from the perspective of complexity Theory. *Emergence: Complexity and Organization, 12*(1).

Morçöl, G. (2012). *A Complexity Theory for Public Policy*. New York: Routledge.

Morin, E. (2005). *Introduction à la pensée complexe*. Paris: Éditions du Seuil.

Moss, S., & Edmonds, B. (2005). Towards Good Social Science. *Journal of Artificial Societies and Social Simulation, 8*(4), 13. Available at http://jasss.soc.surrey.ac.uk/8/4/13.html.

Quéau, P. (1986). *Éloge de la simulation.* Champ Vallon/INA: De la vie des langages à la synthèse des images.

Railsback, S. F., & Grimm, V. (2012). *Agent-Based and Individual-Based Modeling: A Practical Introduction.* Princeton, NJ: Princeton University Press.

Rhodes, D. H., & Ross, A. M. (2010). Five aspects of engineering complex systems emerging constructs and methods. In *2010 IEEE International Systems Conference*, San Diego, CA, pp. 190–195.

Sawyer, K. R. (2005). *Social emergence: Societies as complex systems.* New York: Cambridge University Press.

Siegfried, R. (2014). *Modeling and Simulation of Complex Systems A Framework for Efficient Agent-Based Modeling and Simulation.* Fachmedien, Wiesbaden: Springer.

Simon, H. A. (1962). The Architecture of Complexity. *Proceedings of the American Philosophical Society, 106,* 467–482.

Sokolowski, J. A., & Banks, C. M. (2009). *Modeling and Simulation for Analyzing Global Events.* Hoboken, New Jersey: Wiley.

Sterman, J. D. (2000). *Business Dynamics Systems Thinking and Modeling for a Complex World.* New York: McGraw-Hill.

Treuil, J.-P., Drogoul, A., & Zucker, J.-D. (2008). *Modélisation et simulation à base d'agents: exemples commentés, outils informatiques et questions théoriques.* Paris: Dunod.

Van Dam, K. H., Nikolic, I., & Lukszo, Z. (Eds.). (2013). *Agent-Based Modelling of Socio-Technical Systems.* Dordrecht Heidelberg, New York, London: Springer.

Weisbuch, G., & Zwirn, A. (2010). *Qu'appelle-t-on aujourd'hui les sciences de la complexité?* Vuibert.

Chapter 4
Multiple Streams Theory

A little knowledge that acts is worth infinitely more than much knowledge that is idle.
Khalil Gibran

When he developed the multiple streams theory,[1] Kingdon sought to explain the process of policy development and change and to describe how a problem or a public issue can reach the political agenda, which is the first step in the policy process.

Kingdon (2014) argues that a search for origins of public policies turns out to be futile. Comprehensive, rational policy making is portrayed as impractical for the most part, although there are occasions where it is found. Incrementalism describes parts of the process, particularly the gradual evolution of proposals or policy changes, but does not describe the more discontinuous or sudden agenda change (p. 19).

From this observation, Kingdon was inspired by the Garbage Can Model of decision-making developed by Cohen, March, and Olsen (1972), and he adapted it to the policy development and change process.

According to Lascoumes and Le Galès (2007), Kingdon's work systematized the reflection on the agenda from the perspective of decision dynamics (p. 83).

4.1 Basic Foundations of Multiple Streams Theory

The multiple streams approach focuses on three streams flowing through the system: problem stream, policy or solution stream, and politics stream. They are largely independent of one another, and each develops according to its dynamics and rules (Kingdon, 2014, p. 19).

According to Kingdon (2014), these streams represent the dynamics that structure public and political actions. Kingdon states that they are largely governed by different forces, different considerations, and different styles (p. 88). The possibility of developing or changing a policy is high when these streams are joined or coupled during an opening of a "window of opportunity". In the following paragraphs, we

[1] This theory is also called approach, framework, or model.

A. Guidara, *Policy Decision Modeling with Fuzzy Logic*, Studies in Fuzziness and Soft Computing 405, https://doi.org/10.1007/978-3-030-62628-0_4

review the three streams of the system. Our goal is not to identify the actors that constitute the streams, which is a difficult and risky task, but to explore how this theory can help to shape a conceptual model that represents the process of policy decision emergence.

This review is based on works done by Kingdon (2014), Zahariadis (2007), and Knoepfl, Larrue, Varone, and Savard (2015).

4.1.1 Problem Stream

Kingdon (2014) argues that problem definition and struggles over definition turn out to have important consequences (p. 110). In other words, the definition of a problem requires paying attention to the issue and the available solution. Any public issue is often conceptualized in different ways by various participants and this influences the degree of perception of the problem by the decision-makers.

Kingdon stipulates that problems can come from different sources. They can arise from a systematic analysis of indicators that reveal the presence or the appearance of a problem. The indicators are provided by different sources such as statistics, surveys, and reports produced by public administration or by other organizations. However, indicators are somehow subject to interpretation to determine the nature and the importance of the problem. Kingdon also mentions crises and other significant events that generate a sudden problem. However, he claims that crises and landmark events do not necessarily attract the attention of decision-makers unless they reach a certain level considered as significant or alarming. The crises and landmark events must also affect a certain area or a number considered as significant, and must have a serious impact, real or symbolic.

A significant event can be one that particularly triggers the mobilization of several stakeholders, such as an international crisis or a war. However, the influence of a significant event on the decision process is not certain, especially if the impact of the event is limited in time.

Finally, feedback on existing policies and programs is also a source of issue identification. This feedback can attract decision-makers' attention to the results and consequences of policy because they may indicate the degree of success or failure (Kingdon, 2014). Feedback also comes from evaluation reports and follow-ups or even from informal criticisms.

Without developing further this notion of problem stream, we can conclude that the description demonstrates a vague and ambiguous nature of the notion of problem. At this point, the general description made by Kingdon does not provide aspects that allow this variable to be operationalized. It is possible though to consider the indicators that can be measurable in some cases, but Kingdon claims that their assessment still depends on the interpretation of the decision-makers.

Nevertheless, Knoepfl et al. (2015) propose the following criteria that help to identify some factors of the problem stream.

- The intensity of the problem: This factor is estimated by individual and collective consequences of the problem and by the perception of key actors, as well as the context and the time. The consequences are defined in terms of financial and human costs that stakeholders present to raise awareness or to mobilize the decision-makers and the public.
- The extent of the problem: This factor is closely related to the public visibility given to the issue, and it represents the consequences of the problem on certain groups, the geographical area affected by the problem, and the duration of the consequences. The identification of the area affected requires the knowledge of the population and the regions that are affected or that will be affected by the problem. Besides, a problem may affect a limited area or it can be perceived as "borderless", with a potential risk of slow or rapid spread (Zahariadis, 2007, p. 133).
- The urgency of the problem: The nature of a problem may be considered more or less urgent depending on the context and other considerations. In extreme cases associated with crises, for example, there is potentially an opening of a window of opportunity that mobilizes various audiences and decision-makers.

In addition to these factors, the authors also mention more general factors such as the novelty of the problem that expresses public interest in the problem, although, Zahariadis (2007) claims that new problems and policies are rather rare (p. 134). Knoepfl et al. (2015) also argue that the nature of the problem could also be related to the complexity of the problem, such as a political complexity due to the number of parties involved, or the complexity related to the causes of the problem (single or multiple). But these factors are rather hard to determine.

Despite the imprecise nature of the criteria described above, whether measurable or not, they represent an appropriate manner to determine the sub systemic factors of the problem stream (or the variable Problem). Therefore, we retain the three principal factors for the operationalization of the variable problem and the development of the conceptual model.

4.1.2 Policy Stream

The policy stream is that of the solutions and alternatives. The emergence of alternatives is a selection process in a Policy Primeval Soup, a reservoir of ideas and institutional rules where many ideas float around, bumping into one another, encountering new ideas, and forming combinations and recombinations (Kingdon, 2014, p. 200).

The selection process is not random, and the ideas and solutions promoted follow certain criteria to be included on a shortlist that decision-makers take into consideration. Kingdon argues that these criteria include technical feasibility, congruence with the values of community members, and the anticipation of future constraints, including a budget constraint, public acceptability, and politicians' receptivity

(p. 200). In other words, for a viable proposal to be chosen, it must represent a certain threshold to reach the political agenda.

Kingdon sees this threshold as the result of the survival criteria of a proposal and which requires softening up an effort to make the alternative feasible, acceptable, and perceived as a suitable solution to the problem.

From these details, we can deduce the following factors related to the policy stream (or the variable Solution):

- The suitability of the proposed solution. It indicates whether the solution is appropriate or sometimes unavoidable, even if it does not meet other criteria.
- The feasibility of the solution (operational and technical). Kingdon (2014) argues that without that belief in its technical feasibility, the proposal is not likely to survive to the point of serious consideration (p. 132). This feasibility also includes the financial cost of the solution and its economic impact.
- The acceptability of the solution that represents the compatibility of the proposed alternative with the dominant values. Acceptability is enhanced when the solution is in line with the mainstream thinking of the moment, even if the different groups do not necessarily share the same beliefs and values. Also, acceptability may include the costs and the benefits of the proposed solution.

These factors too are rather vague and uncertain. However, they represent the foundation of the solution stream and therefore they may constitute the sub systemic factors of the variable Solution in the desired conceptual model.

4.1.3 Politics Stream

Regardless of the recognition of the problem and the proposals of solutions, the politics stream evolves according to its own rules and priorities, as advocated by Kingdon. The politics stream is characterized by the following factors:

- The national mood: This factor reflects the state of public opinion or current public trends. The national mood may be favorable or resistant to political initiatives and measures. It is palpable by surveys, debates, and statistics, by implicit support to political choices, and by the refusal of political measures expressed through organized actions against policy choices.
  Kingdon (2014) argues that developments in the politics stream are powerful agenda setters. A new administration, for instance, changes agenda all over town as it highlights its conceptions of problems and its proposals [...] (p. 198).
  It should also be mentioned that popular mood is not simply a reflection of opinion polls, but depends on a whole social, cultural, intellectual, ideological, and political context.
  Moreover, Zahariadis (1999) argues that the politics stream also reflects the ideology and orientations of the partisan groups in power. He specifies that parties

tend to dominate the political stream and exercise considerable control over the shape of policy choices (p. 80).

- Political cohesion and consensus within the government: This factor reflects the absence of conflicts between the various political forces and groups that militate against policy choices. It also represents cohesion within the government. However, the politics stream can also be subject to a form of bargaining within the political power. Kingdon (2014) specifies that consensus is built in the political stream by bargaining more than by persuasion (p. 199). This stream is also dominated by strains between various advocacy and pressure groups that are interested to give direction to the political agenda according to their objectives and interests, and to prevent proposals and policies that are not favorable to their benefits.
- Leeway of political power: This factor depends on the election results and the priorities and prerogatives of the Government. For example, a recent change in political power or administration following an election or an official appointment often gives the government an enhanced leeway. Sometimes, a crisis can also provide greater flexibility to the Government to make choices and impose decisions.

The factors of the politics stream are even more uncertain than those of the two other streams. They represent vague and subjective information. Despite this weakness, we retain these factors of the variable politics since the multiple streams theory does not provide details to further characterize the politics stream.

4.2 Assessment of Multiple-Streams Theory

Multiple streams theory describes a process of policy development and change that results from the coupling of three autonomous streams defined by Kingdon. Each of these streams includes a specific dynamic intended to bring a public issue to the political agenda and triggering policy development or change.

However, this theory does not represent a methodology to deal with complex systems and to simulate the policy decision emergence. Indeed, if we review this theory from the complex system perspective and the requirements of this research project, we can identify the following characteristics that represent the weaknesses of the multiple streams theory for this research:

- Independence of streams and absence of interrelationships: Kingdon points out that the streams are independent of each other and that each stream has largely its own life with its dynamics and rules. These streams do not interact with each other. The coupling of streams, when a window of opportunity is open and with the involvement of political entrepreneurs, represents the only moment when these streams may interact.
- Cause and effect relationships: the approach advocated by Kingdon is linear and deterministic. It suggests that the same conditions lead to the same results and that similar circumstances can produce the same coupling of streams, and maybe

the same decisions and policies. This causal link does not consider the dynamics within the system, which is complex by nature, but this aspect is not explicitly contextualized in the multiple streams theory. We can also deduce from the multiple streams theory that the political agenda depends on the streams' coupling, which in turn relies on the opening of a window of opportunity. This concatenation of events adds a sequential causal link to the process that neglects other dynamics than that of a window opening.

- Order and determinism: the scenarios presumed by multiple streams theory correspond to an ordered and incremental approach in which the overall result represents an addition of the "parts". In other words, when there is a coupling of streams, there is first identification of a problem. Then there is a proposition of a solution or a policy that may be an appropriate response to the problem raised; and finally, there is favorable support of the political power and the public to the proposed solution. This aspect too makes the multiple streams theory a deterministic and linear approach based on an ordered and rational sequential process.
- Systemic and reductionist reasoning: Multiple streams theory reasons at the systemic level to analyze the decision process. It does consider certain variables and actors as well as incrementalism in the decision-making process. However, these elements do not constitute the conceptual setting of Kingdon's approach, which is only based on the streams and their coupling. Zahariadis (2007) confirms that M[ultiple] S[treams] theorizes at the systemic level, and it incorporates an entire system or a separate decision as the unit of analysis (in Sabatier, 2007, p. 66). Furthermore, Zahariadis (2003) argues that […] multiple streams attempts to uncover rationality, theorizing from the macro to the micro (p. 5, 6). This keeps the multiple streams theory far from the requirements of this research project that aims to investigate the dynamics of the micro-level of the decision process and its impact on the policy decision emergence following a bottom-up approach.
- Systemic control of the process: Kingdon (2014) considers the window of opportunity as a trigger for agenda-setting and policy development or change. He claims that it is an opportunity for advocates of proposals to push their pet solutions or to push attention to their special problems (p. 165). He also points out that, often, the various stakeholders stand ready with proposals awaiting problems or with problems awaiting political will or solutions, and that they are watching for the critical moment of opening a window of opportunity.
  Kingdon argues that control of the political agenda by the various stakeholders is the primary way to control the political process. This indicates that it is not the various factors and their interactions that influence the decision process, but rather the participants and entrepreneurs, who lead the political game on their own as if they had control over the whole system and the decision-making environment.

As a matter of conclusion, multiple streams theory is a conventional and systemic policy analysis approach. It aims to restructure past events of the political process, in a top-down manner through linear approximations, to explain the decision-making process, and to identify the circumstances that lead to a window opening and streams coupling. Due to the raised weaknesses of the multiple streams theory, we conclude that it does not constitute, as it is formulated, an appropriate method to deal with the policy decision emergence as a complex system. The complexity properties are absent in the multiple streams theory because it is a positivist and a top-down approach. It investigates the decision as a unit of analysis at the systemic level. While the multiple streams theory proposes some factors that characterize each stream at a deeper level of the political process, it does not address the dynamics between these factors nor their individual impact on the streams and the decision process. However, this approach should not be rejected because it represents the link between complexity and policy decision process. As it was demonstrated, the Stacey Matrix adapted to policy decision-making, indicates the agenda-setting and the Garbage Can Decision-making —two pillars that have shaped the multiple streams theory—as two aspects of the decision process in complex situations (Geyer and Rihani, 2010).

The fundamental notions of complex systems theory, combined with the concepts provided by the Stacey Matrix, allow us to retain the factors that characterize Kingdon's streams as sub systemic or micro-level components for the desired conceptual model of policy decision emergence. We consider that these factors constitute a sub systemic factors of the policy decision emergence whose main components (or variables) are the streams described by Kingdon's theory, i.e. the problem stream, the solution stream, and the politics stream. Even if the multiple streams theory does not satisfy entirely the requirements of this project, it is still a valid basis for building a decision emergence conceptual model. Indeed, a policy is the outcome of a decision, and for a policy decision to emerge, there must be a problem, a solution, and a political will. Figure 4.1 represents the conceptual model of the decision emergence process.

In this conceptual model, it should be made clear that the nature of decision emergence does not correspond to the concept of streams' coupling as defined by Kingdon. This coupling is an addition of phenomena specific to the three independent

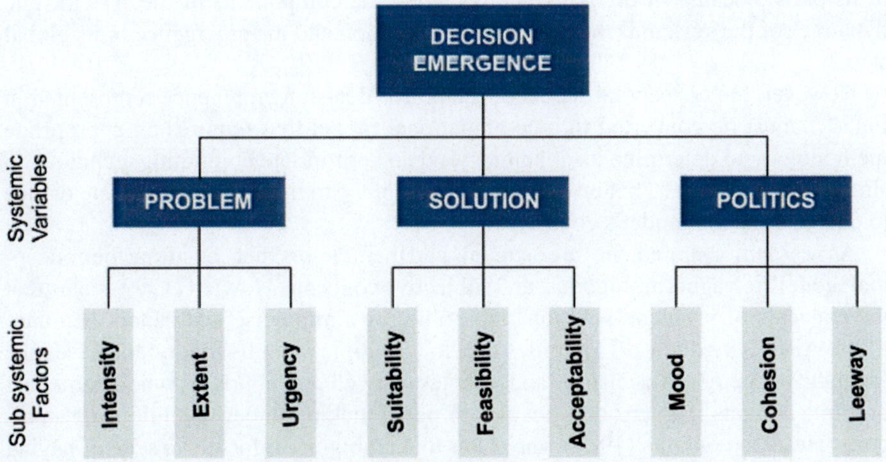

Fig. 4.1 Conceptual model of the decision emergence

streams, and it is conditioned by the opening of a window of opportunity and the action of political entrepreneurs. Therefore, the decision emergence model must not be based on the deterministic concept of coupling, but rather on decision heuristics that are consistent with the policy decision emergence.

This emergence is generated by the dynamics of sub systemic factors of the streams, which differ from the multiple streams settings. The policy decision emergence does not follow the logic defined by Kingdon based on streams' coupling during an opening of a window of opportunity. Therefore, it is necessary to adjust this approach to fit our project's needs, which requires a methodological transition that allows us to simulate the policy decision emergence as a complex system.

4.3 Paradigmatic and Methodological Transition

In the previous sections, we demonstrated that policy decision emergence is a complex system. We also identified the basic factors of the three streams of Kingdon as sub systemic components of that complex system. This statement illustrates, on one hand, the combination of complexity theory and multiple streams theory, and on the other hand, a transition from the systemic (macro-level) decision analysis focused on actors to a sub systemic (micro-level) approach based on factors. This is a paradigmatic shift from the traditional top-down and post-facto analytical approach to a bottom-up and generative approach, specific to complex systems behavior.

The analytical top-down approach is reductionist in the sense that it reduces the system to the sum of its constituent parts. It is also linear because the objective of the analysis is to look for cause and effect relationships. From this perspective, we can also describe this approach as deterministic.

The complex bottom-up approach considers that the whole is more than the sum of its parts, because of the interactions among the components of the system. The dynamics of the system is the source of its evolution and the emergence of its global state.

However, the conceptual model of the policy decision emergence represented in Fig. 4.1 must be converted to a computational model that requires an appropriate methodology to determine its dynamics, and an appropriate computing technique to simulate its behavior. This involves in the first place to make the appropriate choice to characterize the model's components.

As was demonstrated, the three streams and their factors that constitute the conceptual model are vague, ambiguous, and subjective concepts. Howlett (1999) states that the concepts of problem, solution, and politics are imprecise and cannot be quantified. Also, Kingdon (2014) argues that the multiple streams theory addresses the ambiguity inherent in decisions and policies, as well as Zahariadis who claims that multiple streams is a lens that deals with policymaking under conditions of ambiguity. He cites Feldman (1989) who states that ambiguity refers to "a state of having many ways of thinking about the same circumstances or phenomena".

Indeed, the streams and their factors are described by natural language in the form of vague and ambiguous linguistic terms. These terms reflect approximate human reasoning, such as describing a problem as urgent or a solution as suitable, which indicates imprecise and vague information.

Ross (2010) states that the vast majority of the information we have on most processes tends to be nonnumeric and nonalgorithmic. Most of the information is fuzzy and linguistic in form (p. 248).

Furthermore, the sub systemic factors of the model, although some are quantifiable, remain uncertain, because of the assessment due to human reasoning. For example, if the extent of a problem concerns a certain area in a given territory, the assessment of its importance and impact often depends on the decision-makers and his subjective and approximate reasoning. This is even critical when the assessment concerns a decision-making situation, which is by definition ambiguous (March and Olsen, 1985).

Furthermore, ambiguity is one of the sources of the complexity of the decision process, as argued by Ross (2010) who states that complexity in the world generally arises from uncertainty in the form of ambiguity (p. 245). He also claims that to ignore this uncertainty is to ignore the real world, and our understanding of it (p. 246). Ross' comments are supported by Pollitt (2008) who states that, compared to conventional approaches, the characteristics raised above offer an alternative that is more relevant to the real world (p. 127). These statements reinforce our approach to deal with policy decision emergence as a complex system characterized by vagueness, uncertainty, and imprecise human reasoning.

The appropriate approach that addresses situations and decisions under conditions of ambiguity, uncertainty, approximate reasoning, and vague and incomplete information is the fuzzy logic theory. Ross (2010) states that fuzzy logic is a method to formalize the human capacity of imprecise reasoning (p. 117). This branch of mathematics gave rise to the theory of fuzzy sets and fuzzy inference systems. Fuzzy inference systems (FIS) are rule-based systems associated with fuzzy set theory and constitute an artificial intelligence method to represent human reasoning and knowledge.

Ross (2010) claims that the fuzzy rule-based system is most useful in modeling some complex systems that can be observed by humans because they make use of linguistic variables [...]. (Ross, 2010, p. 146). He concludes that for very complex systems where few numerical data exist and where only ambiguous or imprecise information may be available, fuzzy reasoning provides a way to understand system behavior by allowing us to interpolate approximately between observed input and output situations (p. 246). Finally, Ross claimed that fuzzy systems are very useful in [...] situations involving highly complex systems whose behaviors are not well understood [...] (p. 8).

Zgurovsky and Zaychenko (2016) for their part support these arguments, and they claim that there is a new trend in the theory of complex decision-making, which is rapidly developing—making decisions under uncertainty. A promising approach for solving many decision-making problems under uncertainty and incomplete information is based on fuzzy sets and systems theory created by Zadeh (p. 81).

The use of fuzzy logic theory is therefore an obvious choice to deal with the policy decision emergence as a complex system. This choice reflects the second paradigmatic transition concerning the decision-making approaches in the field of public policy. This transition consists of a combination of policy decision modeling, and an artificial intelligence method, namely fuzzy inference systems.

Besides, this second paradigmatic and methodological transition completes the bridging of the approaches raised in this book that connects policy analysis, decision modeling, complex systems, and artificial intelligence.

The following chapter deals with artificial intelligence and fuzzy logic.

References

Cohen, M. D., March J. G., & Olsen J. P. (1972). A Garbage can model of organizational choice. *Administrative Science Quarterly, 17*(1), 1–25.

Feldman, M. S. (1989). *Order without design: Information production and policy making*. Stanford, CA: Stanford University Press.

Geyer, R., & Rihani, S. (2010). *Complexity and public policy: A new approach to twenty-first century politics. Policy and society*. Abingdon: Routledge.

Howlett, M. (1999). Rejoinder to Stuart Soroka, "Policy agenda-setting theory revisited: A critique of Howlett on Downs, Baumgartner and Jones, and Kingdon". *Canadian Journal of Political Science, 32*(4), 773–779.

Kingdon, J. (2014). *Agendas, alternatives and public policies* (2nd ed.). Essex: Pearson Education Limited.

Knoepfl, P., Larrue, C., Varone, F., & Savard, J.-F. (2015). *Analyse et pilotage des politiques publiques*. Québec: Presses de l'Université du Québec.

Lascoumes, P., & Le Galès, P. (2007). *Sociologie de l'action publique*. Paris: Armand Colin.

March, J. G., & Olsen, J.P. (Eds.) (1985). *Ambiguity and choice in organizations*. Oxford University Press.

Pollitt, C. (2008). *Time, policy, management: Governing with the past*. Oxford: Oxford University Press.

Ross, T. J. (2010). *Fuzzy logic. With engineering applications* (3rd Ed.). Wiley.

Sabatier, P. A. (2007). *Theories of the policy process*. Boulder, Colorado: Westview Press.

Zahariadis, N. (1999). Ambiguity, time, and multiple streams. In P. Sabatier (Ed.), *Theories of the policy process* (pp. 73–93) 1st ed. Boulder, CO: Westview.

Zahariadis, N. (2003). *Ambiguity and choice un public policy: Political decision making in modern democracies*. Georgetown University Press.

Zahariadis, N. (2007). The multiple streams framework. Structure, limitations, prospects. In Dans P. A. Sabatier (Dir.), *Theories of the policy process* (p. 65–92). Boulder, Colorado: Westview Press.

Zgurovski, M. Z., Zaychenko, Y. P. (2016). *The fundamentals of computational intelligence: System approach*. Studies in Computational Intelligence (Vol. 652). Switzerland: Springer International Publishing.

Chapter 5
Artificial Intelligence and Fuzzy Logic

As complexity rises, precise statements lose meaning and meaningful statements lose precision.
Albert Einstein

This chapter is a short introduction to artificial intelligence and a synthesis of the theoretical and mathematical foundations of fuzzy logic and fuzzy sets. It includes the concepts needed to understand fuzzy inference systems and to apply this methodology in modeling and simulation of policy decision emergence.

5.1 Artificial and Computational Intelligence

The concept of Artificial Intelligence[1] has appeared in 1956. It refers to a set of knowledge and techniques to deal with complex systems and solving problems that are difficult or impossible to handle using conventional methods, such as decision-making processes. Research in "artificial thinking" began as early as the end of the Second World War, based on the work of Allan Turing and his theory of computation.

In 1956, Allen Newell and Herbert Simon developed a reasoning program called Logic Theorist, which they applied to solve mathematical theorems using search trees with heuristics. Later, other developments have been realized by taking advantage of the algorithmic and computational evolution that occurred afterward.

Wagner-Rémy (2016) suggests that when we are faced with a category of problems in a certain area of expertise and we do not know how to solve it (processing, a sequence of operations or algorithms), but we know what are the laws or rules (knowledge base) that govern this area, we enter the field of "intelligence".

Artificial intelligence is a field of study that cuts across many scientific fields that deal with problems where human reasoning is an important part. Because it is a scientific field, artificial intelligence offers several techniques and methods for dealing with different types of problems and complex systems. As a discipline,

[1]For a full review of the artificial intelligence field and its evolution, see: Flasiński (2016). Introduction to Artificial Intelligence. Springer International Publishing Switzerland. Chapter 1.

A. Guidara, *Policy Decision Modeling with Fuzzy Logic*, Studies in Fuzziness and Soft Computing 405, https://doi.org/10.1007/978-3-030-62628-0_5

the aim of artificial intelligence is therefore to reproduce human reasoning and its ability to evaluate and process complex situations and non-linear phenomena, through theories, techniques, and specific tools.

In the 1990s, Artificial Intelligence (AI) experienced a boom thanks to the progress of computing applications and the increase of the computing power, which enabled the use, the development, and the dissemination of several AI methodologies. This development has affected a wide range of disciplines, such as engineering, social sciences, medicine, linguistics, and telecommunications.

The technological and methodological evolution of the artificial intelligence field gave rise to computational intelligence, an evolving field which consists of the computer application of techniques and methods of artificial intelligence, such as fuzzy inference systems, using modeling and simulation.

5.2 Fuzzy Logic

Fuzzy Logic[2] is a mathematical discipline and an artificial intelligence field. It aims to formulate, understand, and implement processes of approximate reasoning and imprecise or uncertain knowledge described by linguistic variables. Lotfi Zadeh formalized the concept of fuzzy logic in 1965 in his research on the fuzzy set theory,[3] as an extension of binary logic.

Fuzzy logic explores reality beyond the syllogism and limits of binary logic, which was the foundation of positivist Newtonian science. Lucci and Kopec (2016) argue that Fuzzy logic assigns grayness levels to events that were previously declared as black or white (p. 240).

As for vagueness and imprecision, Bouchon-Meunier (2003) states that reasoning in the presence of imprecise knowledge was one of the concerns expressed by L. A. Zadeh, who wanted to approach human reasoning [...]. By reasoning, we mean the general process of using knowledge about a system to build knowledge about that system (p. 121). Bouchon-Meunier also argues that approximate reasoning is the mechanism capable of using and considering imprecise, fuzzy, or uncertain knowledge to produce new knowledge, as human reasoning can do it (p. 29).

Furthermore, Flasiński (2016) claims that fuzzy logic is a very important formalism in the field of artificial intelligence. This theory combines reasoning close to human thought, heuristics, and computing power to process complex systems for which there are no conventional methods of resolution. Moreover, according to Russle and Norvig (2010), fuzzy logic is a response to the difficulty of determining appropriate input variables of systems that are qualified as intelligent (p. 557). These

---

[2]For a historical and mathematical review of fuzzy logic, see: R. Bělohlávek, J. W. Dauben, & G. J. Klir (2017). Fuzzy Logic and Mathematics: A Historical Perspective. New York: Oxford University Press.

[3]See: Zadeh (1965). Fuzzy sets. In *Information and Control*, No. 8, pp. 338–353.

statements support the choice of fuzzy logic as an appropriate methodology to grasp the policy decision emergence as a complex system.

To conclude, we can also specify that the role of fuzzy logic in the resolution of complex systems is not only to establish the complex link between input and output variables (The Mathworks, 2017) but also to discover the internal dynamics of the decision "black box" using linguistic variables and heuristics. These dynamics cannot be explored by conventional decision-making approaches which are based on cause and effect links between the inputs and the outputs of a system using linear and deterministic rules.

5.3 Fuzzy Sets

Fuzzy sets are based on the concept of linguistic variables.[4] A linguistic variable designates words or sentences in natural language to describe hedges or intervals that represent degrees of the truth of a statement depending on human approximate reasoning.

Fuzzy logic introduces the notion of fuzzy sets that generalize classical sets. A fuzzy set contains elements that belong to partial intervals of the set that reflect the varying degrees to which the elements belong within the set. A fuzzy set refers also to a class of elements whose limits are not determined with precision but with nuanced intervals of this class representing partial truths.

The classes of elements are the fuzzy subsets, although they are often described as fuzzy sets and not fuzzy subsets, as specified by Bouchon-Meunier (2007). This author states that the notion of a fuzzy subset allows some degrees in the membership of an element to a class, i.e. to allow an element to belong more or less strongly to this class (p. 7). This nuanced membership results in several degrees of truth or probability that illustrate a form of fuzzy representation of knowledge (Zadeh, 1989). This fuzzy representation that uses vague and imprecise descriptions, is closer to human reasoning and is intended to avoid the arbitrary use of artificially rigid limits of ordinary descriptions, and enables flexibility into characterizations (Bouchon-Meunier, 2007).

As an illustration of the notion of fuzzy subsets, we can take the example of the representation of temperature between two limits (−12 °C, +36 °C) in a given environment. Using linguistic descriptions, the temperature is considered very cold or very hot depending on some criteria. The set of intermediate values between the limits can belong to various intervals called membership functions, described by linguistic variables in the whole set of values called in fuzzy set theory the universe of discourse.

Figure 5.1 shows the concept of the nuanced membership of temperature range and illustrates how measured values are described by words that designate fuzzy

---

[4]For further details, see: L. A. Zadeh (1975). The concept of a linguistic variable and its application to approximate reasoning. Information Sciences. Volume 8, Issue 3, Pages 199–249.

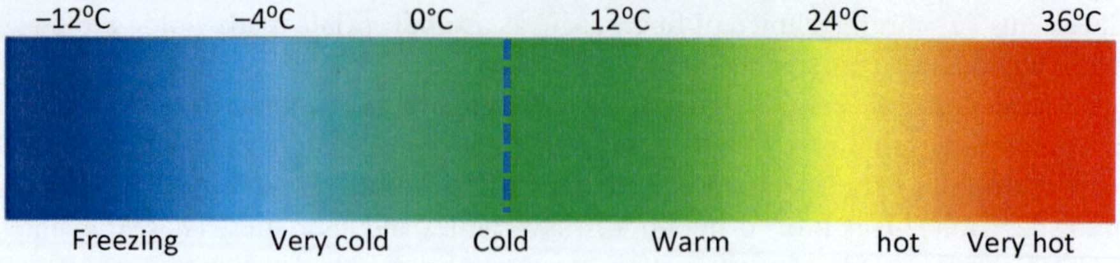

Fig. 5.1 Example of a fuzzy set

states. This example shows how the reasoning goes beyond a simple cold-hot binary representation. It is a kind of partial truth or degree of likelihood of each representative linguistic variable.

In the example above, the linguistic descriptions of temperature values (Freezing to Very hot) constitute the fuzzy variables representing the partial intervals of the universe of discourse. This fuzzy representation illustrates the symbolic mode of reasoning and the way of codifying certain types of information characterized by imprecision and vagueness.

Unlike binary logic in which a statement is false or true, represented by the values 0 or 1, fuzzy logic allows any value in the interval [0, 1] to be assigned to a variable. This is to characterize the partial membership of the variable in the interval or set of values and to allow a more realistic and detailed representation of the graduality. For example, a temperature of 16 °C is described by the subset Warm, which is neither hot nor cold.

Let us elaborate on the concept of partial truth in a certain universe of discourse with the example of water state transformation. It is generally presumed that water transforms from liquid state to solid state (ice) when its temperature (T) is equal to or lower than 0 °C. This statement follows a binary logic in which the transition value is a well-defined constant (curve 1 in Fig. 5.2).

However, this is not always the case since other factors may intervene in the water transformation process, such as the pressure of the environment or the exposure time.

Fig. 5.2 Representation of Binary state (1) and Fuzzy state (2)

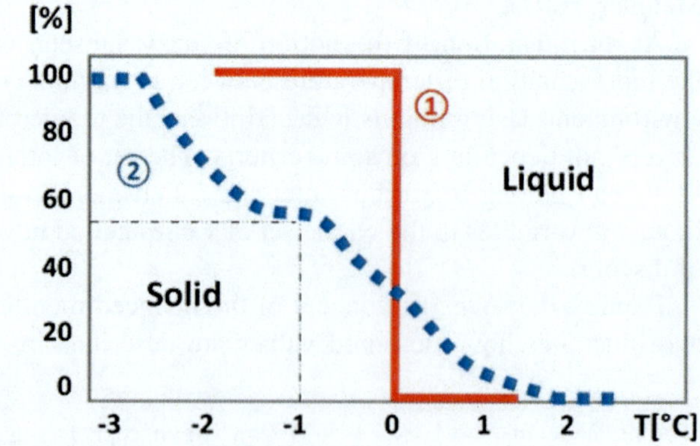

In this case, the water does not necessarily transform into ice at 0 °C, but probably when the temperature is between −3 °C and +2 °C.

In this range, water can be in a gradual mixture of liquid and solid states, to varying degrees; this is a fuzzy state represented by curve 2 in Fig. 5.2.

Curve 2 illustrates the gradual transition between the two states of water. If we take for example the state of the mixture at a temperature of −1 °C, water is neither completely solid nor completely liquid. About 55% of this water is solid and belongs to the solid-state that represents a membership function of the fuzzy set. The rest of the mixture (45%) is liquid and belongs to the liquid state that represents another membership function of the fuzzy set.

5.4 Fuzzy Logic as a Decision Tool

Regarding the use of artificial intelligence (AI) methods for decision processes, Flasiński (2016) states that supporting a process of decision-making was one of the first applications of AI systems (p. 227). The main reason is that decision processes are tainted by uncertainty, and this uncertainty introduces imprecision and ambiguity in the decision process. Sivanandam, Sumathi, and Deepa (2007) argue that the decision is made under risk. When the only available knowledge concerning the outcomes consists of their conditional probability distributions. The uncertainty existing is the prime domain for fuzzy decision (FD) making (p. 151). This statement also supports the choice of fuzzy logic to capture the uncertainty related to decision processes.

It is the uncertainty that influences human reasoning in the processing of available information and in decision making that makes fuzzy logic an appropriate tool for dealing with decision processes. Indeed, fuzzy logic addresses this aspect by introducing certain flexibility in the qualitative evaluation of the variables of a system and in the verification of the rules that govern it. It allows us to consider a certain degree of plausibility, which is closer to reality and more consistent with the very nature of decision processes.

This flexibility has made fuzzy logic an increasingly powerful tool used to solve decision-making problems. However, the implementation of fuzzy logic techniques requires the development of rule-based heuristics by a specialist in the field of application. As a result, the combination of human expertise and algorithmic power makes fuzzy logic a powerful decision-making tool.

Additionally, Shanmuganathan and Samarasinghe (2016) claim that Fuzzy Systems are used to handle inexact data and knowledge in expert systems [...] They are powerful in using inexact, subjective, ambiguous data and vague knowledge elements (p. 12). They add that Fuzzy Systems can represent symbolic knowledge and use numerical representation like the sub-symbolic systems (p. 12), which enhances the power of fuzzy logic to solve various complex problems. The authors take over

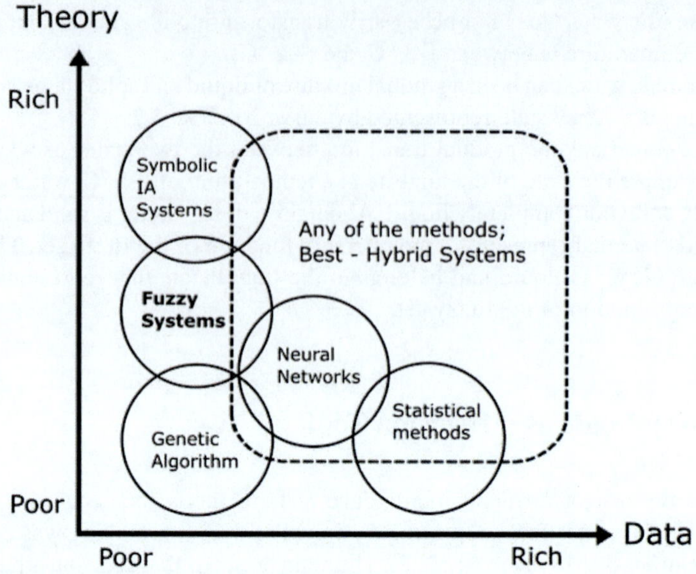

Fig. 5.3 Usability of different methods depending on the availability of data and theoretical expertise. Based on Kasabov (1996, p. 67)

the work of Kasabov[5] who claims that the fuzzy systems technique, derived from the fuzzy logic theory, is recommended especially when the data are incomplete and of low quality, and when the theoretical knowledge concerning the problem under study is not well developed.

Kasabov represents this statement in the following Fig. 5.3.

Flasiński (2016), on his part, argues that if a decision problem is described with fuzzy notions, then fuzzy rule-based systems, […] or hybrid systems based on fuzzy set theory […], can be used (p. 227). This statement is consistent with the work of Kasabov, Shanmuganathan and Samarasinghe, and it strongly supports the choice of the fuzzy sets to deal with the policy decision emerge.

Indeed, the information related to the components of our conceptual model is fuzzy. Therefore, we need to develop a fuzzy inference system that represents the conceptual model of policy decision emergence converted to a computational model. But first, it is necessary to explore the fuzzy inference systems foundations.

[5]N. K. Kasabov (1995). *Foundations of Neural Networks, Fuzzy Systems, and Knowledge Engineering*. A Bradford Book. The MIT Press, Cambridge, Massachusetts, London, England.

5.5 Fuzzy Inference Systems: Principles and Modelling

We dedicate this section to the presentation of fuzzy inference systems[6] (**FIS** hereafter) and their properties. We will also discuss the type of fuzzy inference system chosen for our model, the modeling of fuzzy inference systems, and the computational method for the simulation.

To start this section, let us specify that inference is an operation by which one moves from a first assertion considered true, to a second assertion through a system of rules, that makes the second assertion also true. Inference requires a reasoning subject (Wagner-Rémy, 2016) and allows transforming inputs (or predicates) into outputs (or conclusions) through rules built for this purpose. In the fuzzy set theory, we call these rules fuzzy inference rules.

From an epistemological point of view, Wagner-Rémy (2016) defines inference rules in these terms: the rules that underlie the process of deduction, derivation, or demonstration. Applying the rules to the axioms of the system allows the theorems of the system to be demonstrated. Its arguments are called the "premises" and its value the "conclusion". Rules of inference can also be considered as oriented relationships that link premises to conclusions, and through which a conclusion is said to be "derivable" from the premises (p. 118).

Fuzzy inference rules represent the fuzzy logic type of reasoning and symbolize the relationships between the input variables of the system, described by linguistic variables, and the output variable, also described by a linguistic variable.

Therefore, a fuzzy inference system requires the determination of the input and output variables as well as the rules that determine the relationship between these variables. It is a process of formulating the mapping from a given input to an output using fuzzy logic (The MathWorks, 2017) because the decision process is at the heart of any fuzzy system. Indeed, Sivanandam, Sumathi, and Deepa (2007) argue that decision-making is an important part of the entire system. The FIS formulates suitable rules and based upon the rules the decision is made (p. 118).

5.5.1 *Structure and Modelling of Fuzzy Inference Systems*

As a concept, a fuzzy inference system is a system that converts fuzzified inputs (numerical variables transformed to fuzzy values) to a defuzzified output (fuzzy variable converted to a numerical value) through an algorithm based on inference rules. We represent the structure of a FIS in the following Fig. 5.4.

[6]We use indifferently the terms fuzzy inference systems or fuzzy inference subsystems. Also, in several works, fuzzy inference systems are called rule-based fuzzy systems, fuzzy models, and fuzzy expert systems.

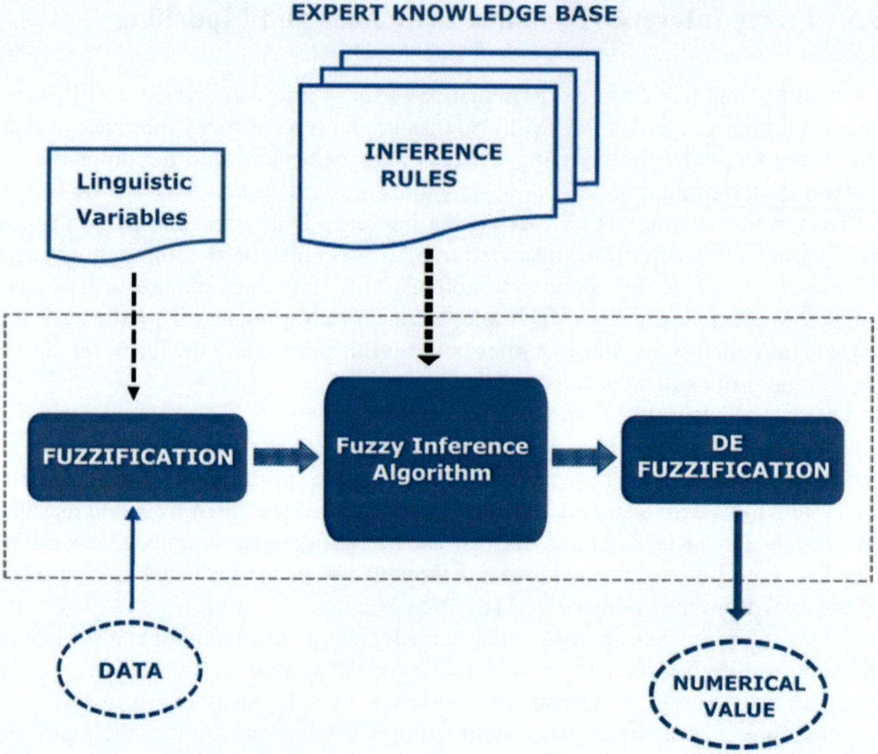

Fig. 5.4 Structure of a fuzzy inference system (FIS)

The structure represented in the Fig. 5.4 also introduces the modeling and simulation steps of a FIS. In this structure, the decision unit is the fuzzy inference algorithm that applies system-specific inference rules during the simulation.

Compared to a conventional system that includes an unknown "black box" between inputs and outputs, FIS demystifies this decisional "black box" through the inference algorithm and the application of decision heuristics or inference rules. For this reason, the FIS is better suited to investigate the decision emergence process. The ability of a FIS to explain what is happening inside the "black box" of the decision process gives the FIS a powerful capacity that is lacking in the conventional analytical approaches.

In the process of modeling and simulation of a FIS, the first step, called fuzzification, consists in transforming numerical data into fuzzy variables and assigning membership functions to these variables according to the expertise of the practitioner and the knowledge base. These membership functions are described by linguistic variables.

The second step is the inference operation performed by the fuzzy inference "engine", which is the algorithm or the decision unit of the system, using fuzzy inference rules defined for this purpose by the practitioner and based on the expertise or the empirical data. The aggregation of the inference rules allows the algorithm to produce a fuzzy output.

Finally, the third step is the defuzzification step and it consists in transforming the fuzzy output variable into a numerical value through the application of defuzzification methods.

The implementation of these steps depends on the nature of the fuzzy subsets, the input variables, and the output variable. The output variable can be a fuzzy variable that is transformed into a numerical value, a constant, or a linear mathematical expression. The choice of the fuzzy inference method, also known as an algorithm or model or inference engine, also depends on the nature of the desired output variable.

5.5.2 Fuzzy Inference Algorithms

Regarding artificial intelligence (AI) and inference systems, Wagner-Rémy (2016) argues that the heart of an AI program is a kind of computing program called an "inference engine", which consists in articulating propositions with each other (p. 19). She adds that such a computing program simulates a form of intelligence, insofar as it does not execute a sequence of operations defined a priori by the programmer, but thanks to its "inference engine", which is responsible for chaining the rules of the knowledge base from a given input situation (base of facts), it arrives at a final output situation, in response to the initial situation (p. 19).

This operation, called mapping, that describes the relationship between input and output variables, is the basis of the development of fuzzy inference models or algorithms. This operation consists in the development of the appropriate algorithm that simulates the variables and the rules to produce a result. The fuzzy inference engine is the heart of any fuzzy inference system (FIS), as it can simulate the "human" decisional thinking that is often based on fuzzy concepts and approximate reasoning.

In fuzzy logic theory, there are several fuzzy inference algorithms, also called fuzzy inference methods or fuzzy inference models. The most popular algorithms are the Mamdani model (sometimes referred to as the Assilian-Mamdani model) and the Sugeno model, referred to as the Takagi Sugeno (TS) model.

The difference between these two algorithms concerns the result produced as the output variable of the system. The Mamdani model produces a fuzzy output variable characterized by a fuzzy membership function, whereas the Sugeno algorithm produces a constant or a linear mathematical expression as the output variable.

Mamdani's algorithm is recommended for dealing with complex and non-linear systems, particularly in multi-criteria decision problems (The MathWorks, 2017), because it has the advantage of including the basis of human knowledge and intuition, through fuzzy inference rules and approximate reasoning.

5.6 Mamdani Fuzzy Model

This model was first designed as a systems control method based on two elements:

- Zadeh's theoretical work on fuzzy algorithms for complex systems and decision-making processes.
- Fuzzy rules from industrial practice.

Mamdani model is considered as the most popular fuzzy inference algorithm because all the existing results on fuzzy systems as universal approximators deal with Mamdani fuzzy systems only, and no result is available for TS fuzzy systems with linear rule consequent (Sivanandam, Sumathi, & Deepa, 2007, p. 119). Even though, our choice is rather dictated by the nature of our project. Indeed, the policy decision emergence is a fuzzy variable and it cannot be represented by a constant value or a linear mathematical expression. Therefore, Mamdani fuzzy model is an obvious choice in this research.

In the following sections, we describe the steps involved in Mamdani fuzzy model represented in the Fig. 5.5.

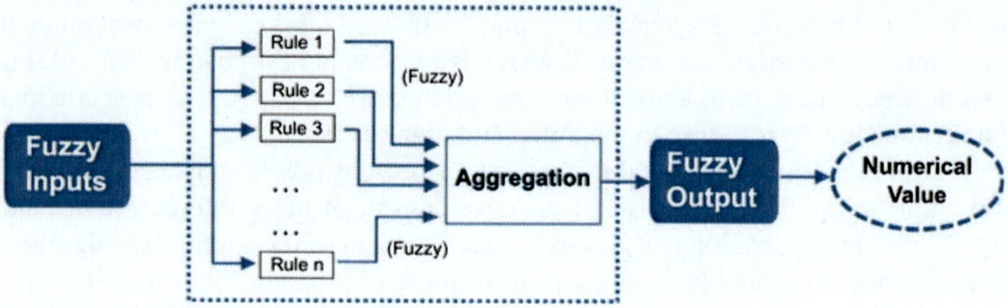

Fig. 5.5 Mamdani fuzzy model

5.6.1 *Fuzzification of Input Variables*

The modeling of a fuzzy inference system starts with a fuzzification step (see Fig. 5.4). This consists in the transformation of an input numerical values into fuzzy variables according to partial intervals of values described by linguistic variables in the range of all values of the system. This requires the determination of the universe of discourse that includes all values and its partition in several fuzzy classes that represent the membership functions.

More specifically, this operation consists of defining the range of all the values that numerical data can have and dividing this range into classes designated by linguistic variables (or fuzzy variables). For example, a set can be divided into three fuzzy classes defined as small, medium, and large. These linguistic variables represent the three subsets in a range from 0% to a maximum of 100% or [0, 1].

To illustrate this technique, let us take the case of an organization that wants to investigate the access rate to an online service that we label **Service Access Rate** (SAR). The objective is to diagnose the factors that explain the tendencies of this access and to monitor the effect of certain factors on this rate. For doing this, let us choose for this SAR a universe of discourse between 0 and 60% or [0, 0.6] partitioned in three fuzzy classes that represent three subsets described by the linguistic variables low, medium, and high, as shown in Table 5.1.

Table. 5.1 Example of membership intervals and fuzzy variables

SAR [%]	Fuzzy variable
0–20	Low
21–40	Medium
41–60	High

SAR: service access rate

Let us also consider that the Service Access Rate (SAR) depends on two variables: **Quality of Service** (QS) and **User Confidence** (UC). To simplify, these variables are each defined by three fuzzy subsets described by the same linguistic variables: low, medium, and high in a universe of discourse also defined by an interval between 0 and 60% like the SAR.

Consequently, we have a fuzzy inference system determined by two input variables (QS and UC) and one output variable (SAR), each having three fuzzy subsets characterized by the linguistic variables determined above. These linguistic variables represent the membership functions.

5.6.2 Membership Functions

Membership Functions are the building blocks of the fuzzy sets, and they are one of the fundamental elements of any fuzzy inference system because any fuzzy set is defined by membership functions (The Mathworks, 2017).

The membership functions can be graphically represented by curves that show the range of variation of the fuzzy subset and the degree to which each input or output value belongs to the subset in the system. These membership functions must be defined for each input and output variable of the inference system in a range between 0 and 100% or [0, 1], and reflect the degree of belonging (membership) of the variable to the defined class or function. In other words, they represent the probability that the numerical value of the variable belongs to a fuzzy subsystem represented by a linguistic variable. This illustrates the way the imprecise information and the approximate reasoning is managed by the fuzzy inference system.

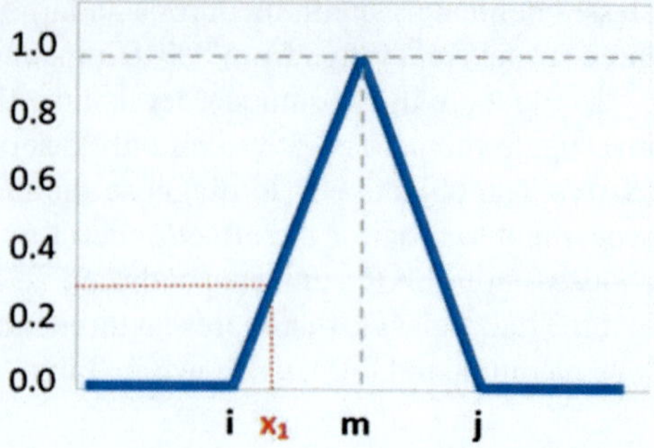

Fig. 5.6 Triangular membership function

There are several types of membership functions: triangular, trapezoidal, Gaussian, sigmoid, etc.[7] These functions are characterized by the shape of their curves and by their calculation formulas. The choice of these functions depends on the model and the application area.

In mathematical terms, let us consider X a set of variables in a fuzzy inference system, and A a fuzzy subset of X. The subset A is characterized by the membership function μ_A having values between 0 and 1:

$$\mu_A : X \to [0, 1]$$

The value of the function $\mu_A(x)$ expresses the degree to which the variable $x \in X$ belongs to the fuzzy subset A.

For example, a triangular membership function is determined by the following formula and represented by the graph in Fig. 5.6.

$$\mu_A(x) = \begin{cases} 0 & x \le i \\ \frac{x-i}{m-i} & i < x \le m \\ \frac{j-x}{j-m} & m < x < j \\ 0 & x \ge j \end{cases}$$

In this example, the m value on the graph represents the maximum membership (100%) of the subset. Any value less than or equal to i and any value greater than or equal to j has a membership function equal to zero (0%).

Any other value in the subset, between i and j, but different from i and j, except m, has a membership function greater than zero and less than 1, provided by the formula $\mu_A(x)$.

As an instance, the value x_1 in Fig. 5.6 has a membership function of 0.3 which means that 30% of it belongs to the subset inside the triangular curve.

[7]For a detailed review of membership functions of fuzzy inference systems, see: S. N. Sivanandam, S. Sumathi and S. N. Deepa (2007). Introduction to Fuzzy Logic using MATLAB, Chap. 4. Springer-Verlag Berlin Heidelberg.

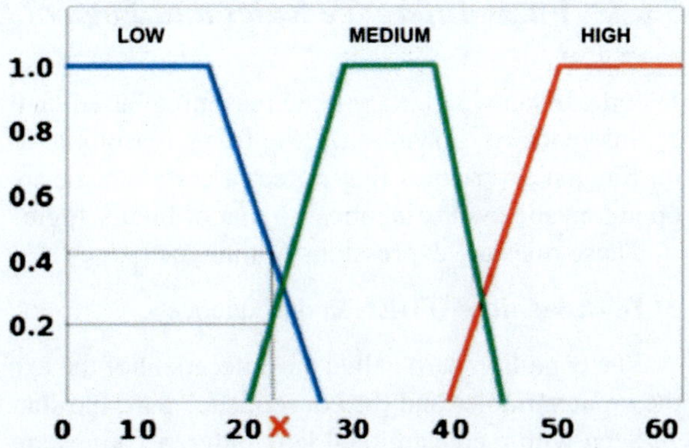

Fig. 5.7 Trapezoidal membership function representing three fuzzy sets

Let us go back to the example of the Service Access Rate (SAR), we defined each system variable as a universe of discourse between 0 and 60% partitioned into three fuzzy subsets that represent three membership functions: Low, medium, and high. Figure 5.7 represents trapezoidal membership functions that apply to the SAR example.

In Fig. 5.7, each trapezoidal curve represents a fuzzy subset designated by a linguistic variable (low, medium, high). Any input value in the universe of discourse (0–60%) belongs to one or more fuzzy subsets to varying degrees.

For example, the x value in Fig. 5.7 has a membership function low equal to 0.45 (or 45%) and a membership function medium equal to 0.2 (or 20%). The value x belongs to both intervals at varying degrees.

In terms of mathematical relations:

$$\mu_{Low}(x) = 0.45 \quad \text{and} \quad \mu_{Medium}(x) = 0.2$$

This membership spread over several subsets illustrates the imprecision and reflects the approximate reasoning regarding the value of the variable x.

To conclude with the membership functions, let us specify that there are several manners to determine the appropriate type of membership functions because there are no theoretical foundations to determine their suitability with the problem under study. In general, the appropriate type of membership function is mainly determined by estimates based on expert knowledge in the field of application. It can also be determined through an exploration of empirical data of the problem, and by trial and error. However, the choice of the appropriate type of membership functions is important for the overall performance of any fuzzy inference models. Therefore, an experimental approach of different types of membership functions is generally needed to get conclusive results.

5.6.3 Fuzzy Inference Rules and Logical Operators

In comparison with binary logic reasoning based on the notions of false or true (0 or 1), inference rules symbolize the fuzzy reasoning. They represent heuristics based on linguistic variables that reflect a certain scale in natural language while using operators and syntax identical to that of binary logic.

These rules are expressions of the type:

IF <condition> **THEN** <consequence>.

The condition part, called the antecedent of the expression, represents the state of the input variables and the consequence part, the state of the output variable.

Seen with a conventional lens, inference rules are heuristics or assumptions for projecting the impact of fuzzy input variables on the system output variable which is the consequence. The output variable of a Mamdani model is a fuzzy variable with its membership functions, as in our example of Service Access Rate (SAR).

Since the antecedent of the expression can be a combination of several conditions related to input variables that represent plausible situations, fuzzy inference rules are developed with the support of the binary logic operators AND, OR, NOT. This way, the rules indicate the combined effect of the input fuzzy variables on the output variable.

According to Zadeh (1965), the logical operator AND designates the intersection of two fuzzy subsets A and B (for example). Therefore, it represents the minimum value as a result of the intersection. In contrast, the operator OR designates the union of two fuzzy subsets A and B and expresses the maximum value that results from this union. Finally, the operator NOT represents the negation or the logical complement of a fuzzy subset. As an example, Fig. 5.8 represents a complement of a set A in a universe of discourse U, such as if:

$$\mathrm{A} = 0.4(\text{or } 40\%),\ \text{the complement of A } (\overline{\mathrm{A}} \text{ or NOT A}) \text{ is}$$
$$\overline{\mathbf{A}} = (\mathbf{1} - \mathbf{A}) = 0.6 \ (\text{or } 60\%)$$

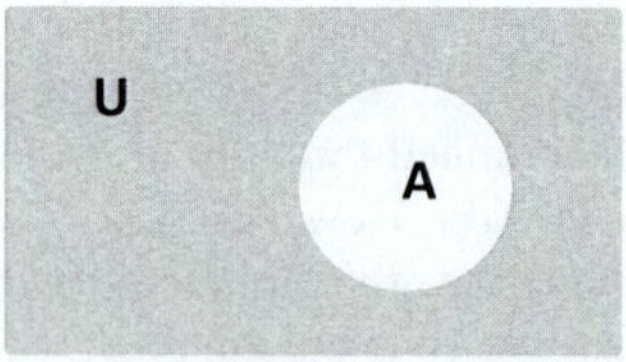

Fig. 5.8 Complement of a subset A (Operator NOT)

The operations carried out on the membership functions μ_A and μ_B (in Sect. 5.6.2) with the logical operators are represented by the following formulas (Table 5.2):

Table. 5.2 Logical and fuzzy operators

Logical operator	Operation	Logical operator (Zadeh)
AND	Intersection	$\mu_{A \cap B} = \min(\mu_A, \mu_B)$
OR	Union	$\mu_{A \cup B} = \max(\mu_A, \mu_B)$
NOT	Complement	$\mu_{\bar{A}} = 1 - \mu_A$

Let us illustrate with an example represented in Fig. 5.9. For two input values x and y belonging respectively to two subsets A and B, such as $x \in A$ and $y \in B$. Their respective membership functions are $\mu_A(x) = 0.6$ and $\mu_B(y) = 0.35$.

Applying the fuzzy operator AND in the <condition> part gives a result of a membership function equal to 0.35 in the <consequence> part because the operator results in the minimum of the two values.

For the same values, the application of the operators OR gives a result of 0.6.

In a fuzzy inference model, the determination of the inference rules is an essential part, since it is the set of fuzzy rules that determines the quality of the inference and the output variable, which represents the result given by application or aggregation of the inference rules. Also, in any fuzzy inference system, it takes as many rules as there are possible presumed scenarios. In general, this is based on the practitioner's expertise in the field.

To make sure to cover all scenarios, a decision matrix in the form of a truth table is necessary to determine all the inference rules of a system. The table includes the fuzzy values of the input variables and the rule in the <condition> part, and the expected output of the scenario in the <consequence> part, which is the outcome provided by the inference rules. Then, the truth table is converted into a set of rules that include all the possible scenarios.

To illustrate this technique, let us apply these notions to our example of Service Access Rate (SAR) defined with the two input fuzzy variables Quality of Service (QS), and User Confidence (UC). For this, we can formulate a set of rules with the operator AND such as:

Rule 1: **IF** QS is low **AND** UC is medium **THEN** SAR is low.

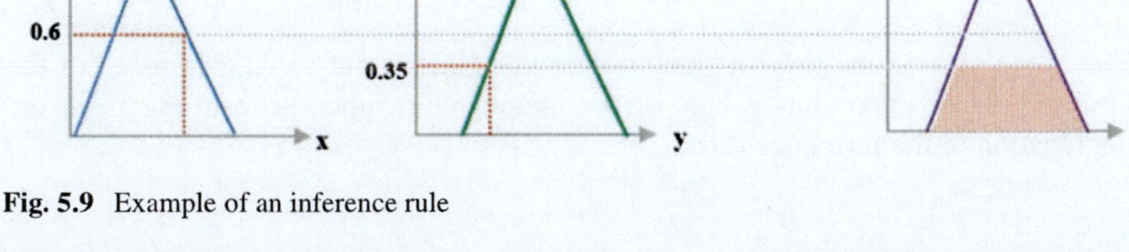

Fig. 5.9 Example of an inference rule

Rule 2: **IF** QS is high **AND** UC is medium **THEN** SAR is medium.
Rule 3: **IF** QS is medium **AND** UC is high **THEN** SAR is high.
Rule 4: **IF** QS is low **AND** UC is high **THEN** SAR is low.
Etc.

The practitioner determines as many "expert" rules as necessary according to the plausible or predicted scenarios.

5.6.4 Defuzzification

The last step in the Mamdani fuzzy model is the defuzzification that converts the output fuzzy variable, given by aggregation of the fuzzy inference rules, to a numerical value. Regarding the aggregation of the inference rules, it should be noted that since decisions are based on the testing of all the fuzzy inference rules of the system, the rules must be combined in such a way to generate a decision. Aggregation is the process by which the fuzzy sets that represent the outputs of each rule are combined into a single fuzzy set (The MathWorks, 2017, pp. 2–25).

In the defuzzification step, the Mamdani algorithm generates a fuzzy output result in the form of a membership function. This fuzzy result is then converted to a numerical value. The conversion is performed using defuzzification methods, such as the Center of Gravity (COG) method, and the Mean of Maximum (MOM) method, which are the most popular methods according to several sources. Let us explain how these methods work.

5.6.4.1 Center of Gravity Defuzzification Method

The Center of Gravity (COG) method consists of calculating the position (in abscissa) of the center of gravity of the membership function resulting from the aggregation of the inference rules. This position represents the numerical value of the output variable. This method is very popular and it is generally preferable because of its consistency with the principles of fuzzy logic, as it incorporates the notion of belonging of a variable to two subsets at the same time. However, it is a method that requires complex calculations and a high computing power especially when the membership functions are not "linear" like triangular and trapezoidal membership functions.

5.6.4.2 Mean of Maxima Defuzzification Method

The Mean of Maxima (MOM) method provides the output value as a mean of the abscissa of the maximum values of the membership function resulting from the aggregation of the inference rules.

Table. 5.3 Principal defuzzification methods

Defuzzification method	Description	Formula (Output value S)	Graphical representation
Center of Gravity (COG) or centroid	Abscissa corresponding to the center of gravity of the surface of the membership function resulting from aggregation	$S = \frac{\sum_{i=0}^{N} X_i.\mu_{xi}}{\sum_{i=0}^{N} \mu_{xi}}$	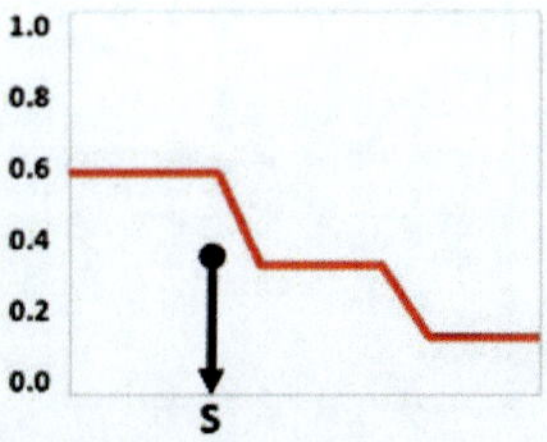
Mean of Maxima (MOM)	Abscissa corresponding to the average of the abscissae with the maximum values of the membership functions resulting from the aggregation	$S = \frac{\sum x_i}{N}$	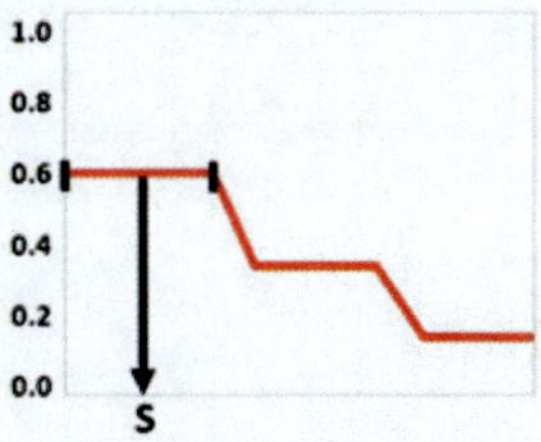

N: Set of values X_i.

Table 5.3 shows the mathematical relations for calculating the numerical output value (S) and the graphical representation of each method. In this table, N is the set of values of the membership functions resulting from the aggregation of the inference rules.

Let us return to the example of Service Access rate (SAR) and apply these defuzzification methods. Let us take hypothetical percentage values for the two input variables (QS = 36% and UC = 44%) in a universe of discourse of 60% for all variables. Let us also chose as a set of rules the first three inference rules mentioned in Sect. 5.6.3 to do the exercise.

The simulation performed with Fuzzy Logic Toolbox™ using these values of QS and UC combined with this set of chosen rules and a COG defuzzification method provides the following result as represented in Fig. 5.10:

$$\mathbf{SAR = 50.6}$$

This result obtained in this example is obviously incomplete and hypothetical, as a comprehensive set of valid rules is not applied for this simulation. Also, the simulation is only done with two basic subsystems (variables) that do not include any sub systemic factors to determine their real scale and impact.

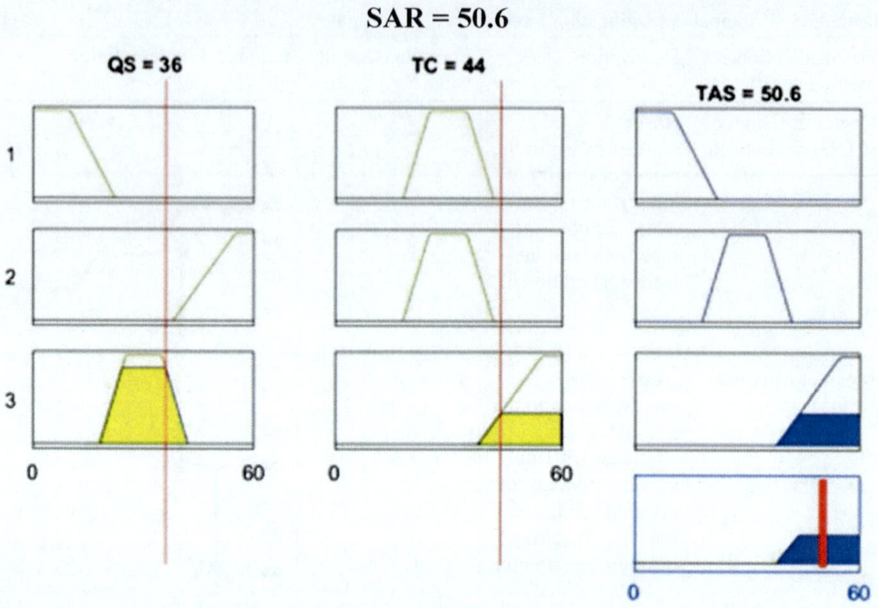

Fig. 5.10 Example of simulation and defuzzification (using Fuzzy Logic Toolbox™)

5.6.5 Modeling and Simulation Tools

In this section, we present the last aspect related to the fuzzy inference systems that makes possible the simulation of the policy decision emergence. This aspect deals with Matlab® Fuzzy Logic Toolbox™ as a computational method. This toolbox is one of the most popular and powerful computing tools for modeling and simulating fuzzy inference models. However, the methodology developed in this book does not depend on this specific toolbox. The are many other computing techniques dedicated to fuzzy logic and fuzzy inference systems.[8] Furthermore, the model developed in this research using Fuzzy Logic Toolbox™ can also be converted to a Python library with a built-in module that provides access to the system's functionalities without the need of Matlab®.

According to MathWorks (2017), Fuzzy Logic Toolbox™ lets you model complex system behaviors using simple logic rules, and then implement these rules in a fuzzy inference system (pp. 1–2). It includes several interrelated complementary tools dedicated to the modeling and simulation steps of the fuzzy inference model. Figure 5.11 shows the tools of the Fuzzy Logic Toolbox™ dedicated to the design of fuzzy inference systems.

[8]See: Nhivekar G. S., Nirmale S. S. and Mudholkar R. R. (2012). A Survey of Fuzzy Logic Tools for Fuzzy-based System Design, International Conference in Recent Trends in Information Technology and Computer Science (ICRTITCS - 2012), Proceedings published in International Journal of Computer Applications® (IJCA) (0975 - 8887), pp. 25–28.

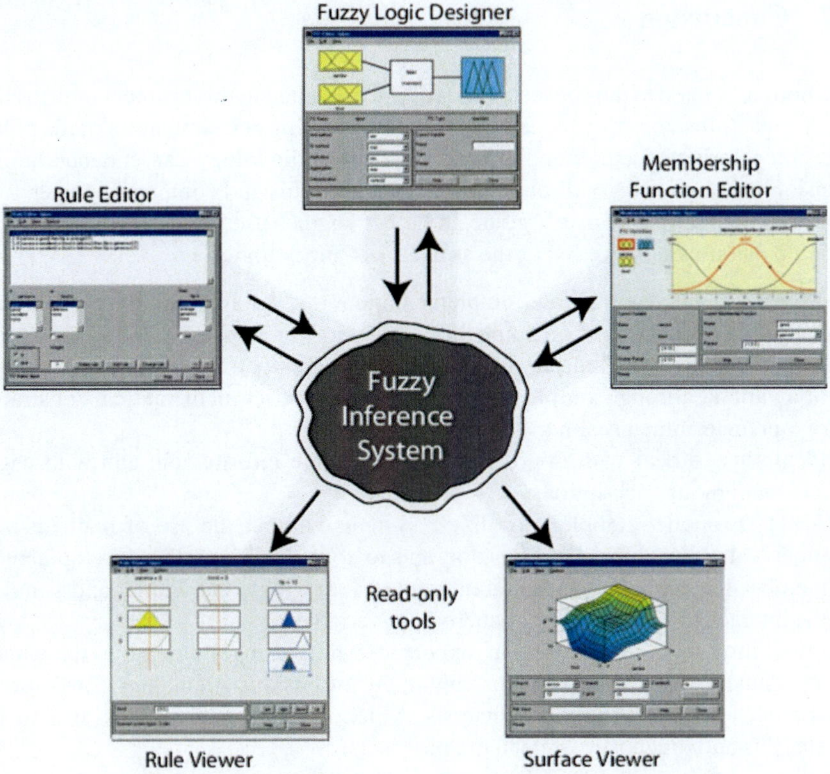

Fig. 5.11 Fuzzy logic toolbox™ modules *Source* The Mathworks, Matlab® user's guide (2017)

Table 5.4 describes the various modules of Fuzzy Logic Toolbox® represented in Fig. 5.11 and their functions.

In Table 5.4, we can see the concepts presented in this chapter included in Fuzzy Logic Toolbox™. The process of modeling and simulation of a fuzzy inference model requires one or more tools of this toolbox.

Table. 5.4 Functions of fuzzy logic toolbox™ module

Modules	Functions
Fuzzy logic designer	Determination of input and output variables
Membership function editor	Definition of membership functions for each variable
Rule editor	Determination of system inference rules
Rule viewer	Visualization of the results of each rule, and the output of the FIS It also shows how the shape of certain membership functions influences the overall result
Surface viewer	Graphical representation of the results

5.7 Conclusion

The notions related to the chosen methodology presented in this chapter confirms that fuzzy logic is the appropriate methodology for modeling and simulation of the policy decision emergence as a complex system. This methodology can enhance human decision-making capacity through advanced algorithms and computing power.

Based on the Matlab user's guide and other sources used in this chapter, we can draw up a non-exhaustive list of the strengths of fuzzy logic:

- Ease of understanding and implementation due to its proximity to human reasoning and its way of grasping the complexity of real phenomena while relying on natural language. Indeed, fuzzy logic allows greater freedom of representation of a variable through simple notions and without strict mathematical constraints or a predetermined scaling.
- Flexibility to deal with imprecise and incomplete information, and with vague and ambiguous descriptions.
- Ability to model complex non-linear systems without the use of mathematical models that are difficult to develop and to solve. Through the development of membership functions and inference rules, fuzzy logic can model and simulate the interactions between input and output variables.
- Possibility to consider human expertise while using computing techniques to simulate that expertise. This might, by extension, strengthen conventional approaches of public policy analysis, which represents an added value to the analysis and diagnostic of decision processes.

Besides, while some aspects of fuzzy logic appear simple, this method represents a powerful tool of artificial intelligence to deal with complex systems. Labiod and Beylot (2013) summarize the main strengths of fuzzy logic by stating that fuzzy logic is considered as a theory for dealing with uncertainty about a complex system, and as an approximation theory. Fuzzy logic has two objectives: on the one hand, to develop computational methods that can perform reasoning and problem solving that require human intelligence, and on the other hand, to explore an efficient way and a trade-off between accuracy and cost in developing an approximate model of a complex system (pp. 184, 185).

Fuzzy logic is far from being a simple theory. It has a strong mathematical foundation, as Ross points out. He claims that just as an algebraic function maps an input variable to an output variable, a fuzzy system maps an input group to an output group; in the latter, these groups can be linguistic propositions or other forms of fuzzy information. The foundation on which fuzzy systems theory rests is a fundamental theorem from real analysis in algebra known as the Stone–Weierstrass theorem, first developed in the late nineteenth century by Weierstrass (1885), then simplified by Stone (1937) (Ross, 2010, p. 7).

However, fuzzy logic also has a weakness that stems from its very concept, as it relies in part on human reasoning, which is never completely free from shortcomings and subjectivity.

Although the designer's choice of variables and rules of a fuzzy inference model can be revised and corrected during the simulations, there is no guarantee that this choice is the best one or that it accurately represents the real aspects treated by the model. Such subjectivity requires rigor and a sense of reality. As Matlab user's guide states, Fuzzy logic is the codification of common sense—use common sense when you implement it and you will probably make the right decision.

However, Ross (2010) reminds that in his Republic (360 BC), Plato suggests the idea that things that are perceived are only imperfect copies of the true reality that can only be comprehended by pure thought (p. 7).

In terms of decision-making, these reflections take us to the assertions of Newell and Simon (1972) regarding artificial intelligence. These authors already argued that a comprehensive understanding of human decision making would be required if A[rtificial] I[ntelligence] was to yield substantial benefits (in Pomerol & Adam, 2008, p. 3).

References

Bouchon-Meunier, B. (2003). *Logique floue, principes, aide à la décision*. Paris: Hermès Science Publications.

Bouchon-Meunier, B. (2007). *La logique floue* (4e edition). Paris: Presses Universitaires de France.

Flasiński, M. (2016). *Introduction to artificial intelligence*. Springer International Publishing Switzerland.

Kasabov, N. K. (1996). Foundations of neural networks, fuzzy systems, and knowledge engineering. In *A Bradford book*. Massachusetts London, England: The MIT Press Cambridge.

Labiod, H., & Beylot, A.-L. (2013). Réseaux véhiculaires, modèles et algorithmes. Paris. Lavoisier.

Lucci, S., & Kopec, D. (2016). *Artificial intelligence in the 21st century: A living introduction 2/E*. Mercury Learning and Information, Dulles, VA.

Newell, A., & Simon, H. A. (1972). *Human Problem Solving*. Englewood Cliffs, NJ: Prentice-Hall.

Pomerol, J-C. & Adam, F. (2008). Understanding human decision making—A fundamental Step towards effective intelligent decision support chapter. *Intelligent decision making: An AI based Approach (Vol. 97). Studies in computational intelligence* pp. 3–40. Berlin Heidelberg: Springer-Verlag.

Ross, T. J. (2010). *Fuzzy logic. With engineering applications* (3rd ed.). Hoboken: Wiley Ltd.

Russel, S., & Norvig, P. (2010). *Artificial intelligence. A modern approach* (3rd ed.). Upper Saddle River, New Jersey: Pearson Education Inc.

Shanmuganathan, S., & Samarasinghe, S. (Eds.). (2016). *Artificial neural network modelling*. Cham: Springer International Publishing Switzerland.

Sivanandam, S. N., Sumathi, S., & Deepa, S. N. (2007). *Introduction to fuzzy logic using MATLAB*. Berlin Heidelberg: Springer.

The MathWorks (2017). MATLAB user's guide. Natick, MA: The MathWorks, Inc.

Wagner-Rémy, C. (2016). *La pensée dirigée: Traité sur le raisonnement et les logiques*. Edition: BoD – Books on Demand, Paris.

Zadeh, L. A. (1965). Fuzzy sets. *Information and Control, 8*, 338–353.

Zadeh, L. A. (1989). Knowledge representation in fuzzy logic. *IEEE Transactions on Knowledge and Data Engineering, 1*(1), 253–283.

Chapter 6
PODESIM—Policy Decision Emergence Simulation Model

The purpose of models is not to fit the data but to sharpen the questions.
Samuel Karlin

This chapter is dedicated to the implementation of the diverse notions developed in the previous chapters. It also describes the major contribution of this research project: the development of the experimental Policy Decision Emergence Simulation Model, which we refer to in the rest of the book by the acronym PODESIM. This experimental model is a decision diagnostic tool whose objective is to identify the levers of the policy decision emergence, that occurs in the sub systemic environment of the decision process, qualified as a complex system.

In this development, the design of the model represents an innovation in public policy studies that converts a linear estimating approach to a simulation model. It consists in transforming Kingdon's multiple streams theory, a public policy narrative approach, into a computational model, through the operationalization of specific variables and factors. The transformation of the multiple streams theoretical concepts into a computational model results in a fuzzy inference modular system, a multi-level simulation model using Fuzzy Logic Toolbox™.

In Chap. 4, we concluded that the three streams of Kingdon's theory are aggregates that represent the main variables of the policy decision emergence. These variables are transformed into three fuzzy inference subsystems (FIS) that constitute the core of the PODESIM model. Each FIS is characterized by factors that constitute the input data to each subsystem. Figure 6.1 represents the modular structure of the PODESIM model that includes the multi-level fuzzy systems and components. This model is designed with Fuzzy Logic Designer, one of the tools of the Fuzzy Logic Toolbox™.

More specifically, the PODESIM model is the transformation of the conceptual model presented in Fig. 4.1 in a computational fuzzy inference model in Fuzzy Logic Toolbox™ to perform the necessary simulation.

The multi-level modular structure in Fig. 6.1 includes three levels and each level constitutes an input to a higher level:

A. Guidara, *Policy Decision Modeling with Fuzzy Logic*, Studies in Fuzziness and Soft Computing 405, https://doi.org/10.1007/978-3-030-62628-0_6

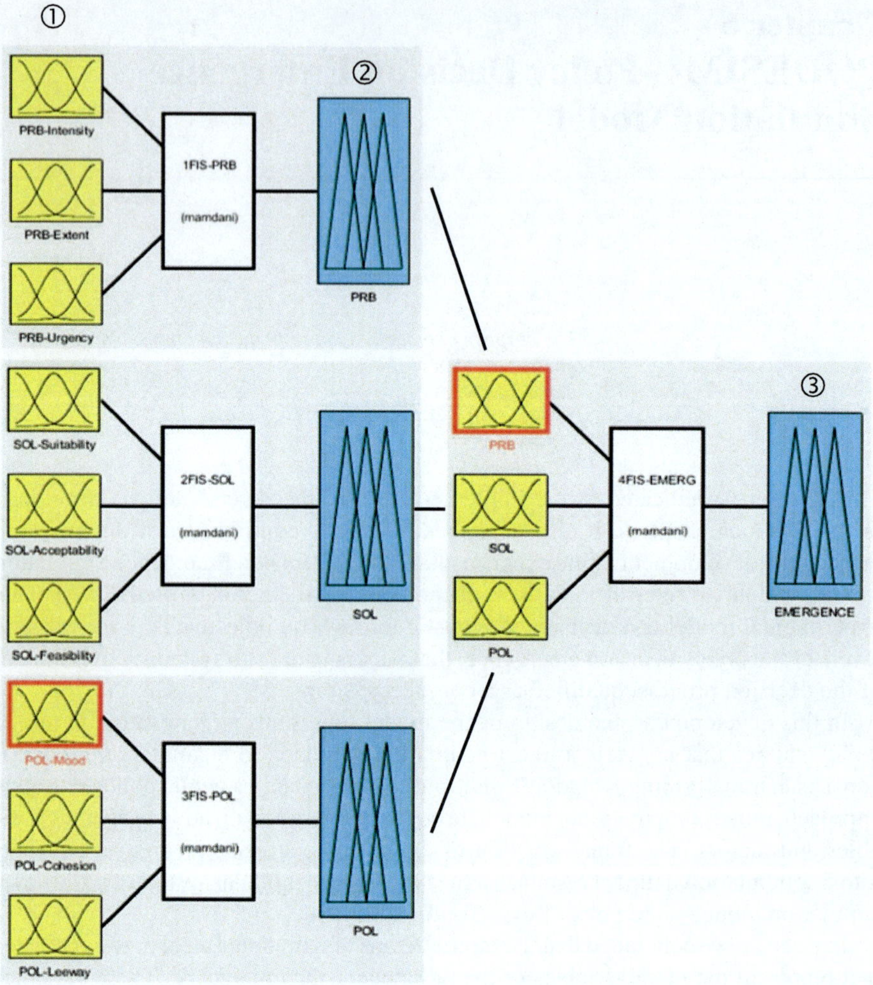

Fig. 6.1 PODESIM fuzzy model

- The lowest level ① includes the sub systemic fuzzy factors of each fuzzy inference subsystems (FIS) of the model (Problem, Solution, and Politics). These fuzzy factors are the input values for each FIS.
- The medium level ② includes the three fuzzy inference subsystems of the model: Problem (PRB), Solution (SOL), and Politics (POL). The output values of these FISs constitute the inputs to the highest level.
- The highest level ③ is the level of policy decision emergence. The FIS Emergence (EMERG) converts the fuzzy result given by the simulation into a numerical value as the output of the system that represents the scale of the decision emergence.

Each fuzzy factor and each fuzzy inference system of PODESIM is represented by its membership functions that describe the ranges of values (or the universe of discourse) of each component of the system.

6.1 Membership Functions of PODESIM

As we explained, the determination of the membership functions and their type is a very important task in fuzzy system modeling. Different techniques exist for this purpose depending on the field of application. However, it is usually a set of data such as time series that allows to define a universe of discourse and to divide it into several intervals described by linguistic (or fuzzy) variables that represent membership functions.

For this project, we do not have data to determine the membership functions for the components of PODESIM. The model is only created out from theoretical concepts and does not rely on any empirical basis so far. Therefore, only the nature of the components and the application area can guide us to identify a suitable universe of discourse for the component and to proceed with the simulation.

Since PODESIM deals with the policy decision field, we used our knowledge in the field of public policy and our judgment to determine a basic, but consistent, universe of discourse for the components.

We choose a universe of discourse between 0% as a minimum value and 75% as a maximum value for all factors and variables, except for the variable Emergence. We justified this choice by hypothesizing that most of the components of the model very rarely exceed a maximum percentage of 75%, for example, the acceptability of a solution or the extent of a problem. Our choice is also motivated by the case study chosen for this research project. This case concerns a serious international crisis that commands a large universe of discourse.

We divided the universe of discourse of all components into three intervals, each characterized by a membership function designated by a linguistic (fuzzy) variable represented in Table 6.1.

Table 6.1 Partition of the universe of discourse of factors and variables of PODESIM

Interval [%]	Membership range
0–25	Low
26–50	Medium
51–75	High

The next step consists in choosing the type of membership functions. For this task, we also base our decision on the judgment because we do not have an empirical background to make an evidence-based choice. Therefore, we choose two types of popular[1] membership functions, the trapezoidal and the Gaussian membership types. In this research, we intentionally excluded the triangular type of membership functions because of the high linearity of this type of function.

In practice, trial and error are often necessary to optimize the choice and the quality of the membership functions because of the several techniques available. MathWorks (2017) suggests that the function itself can be an arbitrary curve whose shape we can define as one that suits us in terms of simplicity, convenience, speed, and efficiency (pp. 2–25, 2–26). This topic remains an investigation area that requires further research and development, especially in the policy decision field.

To illustrate our choice, the following figures show the (trapezoidal) membership functions of the Fuzzy Inference Systems (FIS) of PODESIM model. These figures are generated with Matlab Fuzzy Logic designer and they describe the factors of each FIS, represented by three membership functions.

In Fig. 6.2a, we represent the Problem FIS (PRB) that comprises three specific fuzzy factors: Problem Intensity (PRB-Intensity), Problem Extent (PRB-Extent), and Problem Urgency (PRB-Urgency). The other FISs of the model are represented in the same way, with their respective factors and membership functions as shown in Fig. 6.2b, c.

For the system's output variable, the FIS Emergence, we chose four membership functions. The purpose of this choice is to obtain a higher accuracy of the needed outcome of the system, the decision emergence. These membership functions are represented by the following linguistic variables, describing four intervals in a universe of discourse ranging from 0 to 60% and more. These intervals are:

- Low (0–20%),
- Medium (21–40%),
- High (41–60%),
- Urgent (+60%).

Figure 6.3a, b represent the Decision Emergence FIS with two types of membership functions (trapezoidal and Gaussian, respectively) and include the input fuzzy factors (FISs) which are Problem (PRB), Solution (SOL), and Politics (POL). We remind that the outputs of these medium-level FISs become the inputs to the highest-level Fuzzy Inference System, the FIS Emergence.

We experiment in this project both types of membership functions above.

[1]For more details on this topic, see: Bouchon-Meunier, Dotoli, & Maione, (2007).

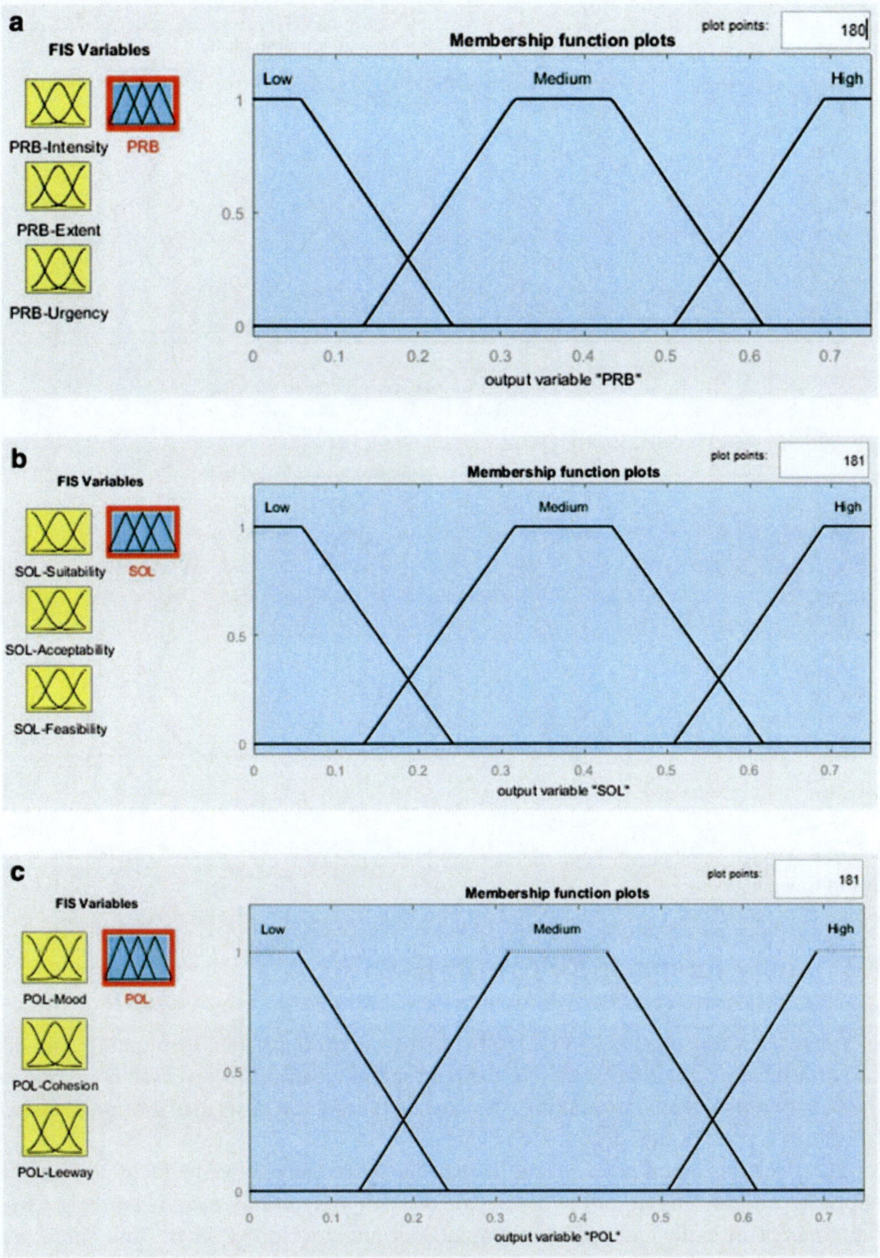

Fig. 6.2 **a** Trapezoidal membership functions of FIS Problem (PRB), **b** trapezoidal membership functions of FIS Solution (SOL), **c** trapezoidal membership functions of FIS politics (POL)

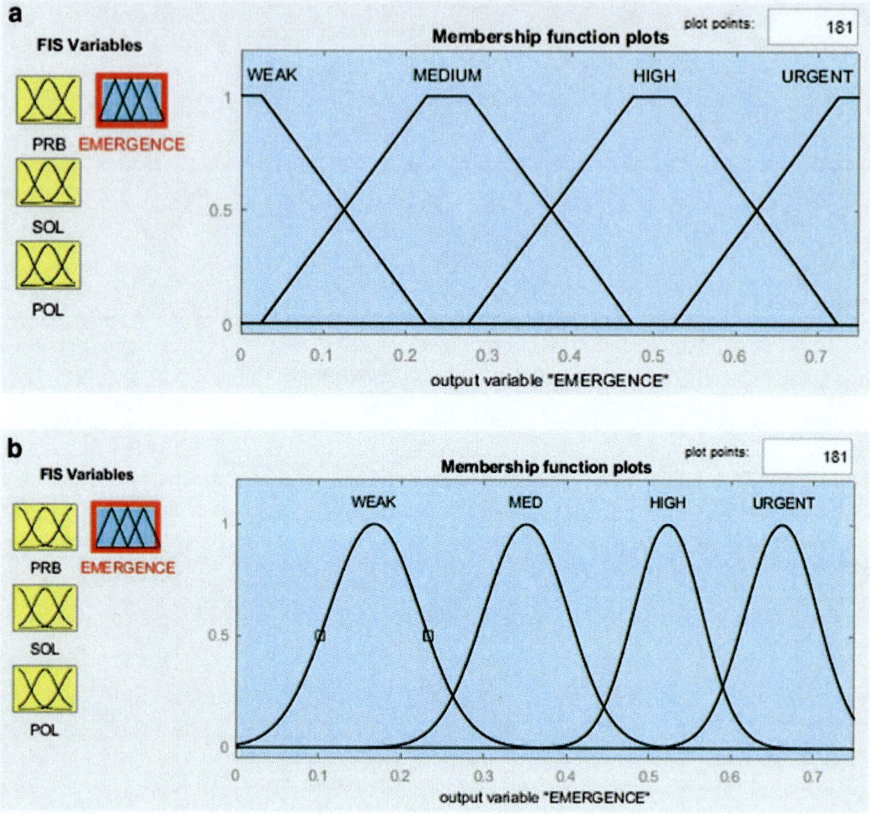

Fig. 6.3 **a** Trapezoidal membership function of FIS Emergence, **b** Gaussian membership function of FIS Emergence

6.2 Fuzzy Inference Rules of PODESIM

The fuzzy inference rules of PODESIM are decision heuristics formulated specifically for the case study presented in the next section. Since this research is unique in the field of policy decision-making, we have no theoretical basis to elaborate decision heuristics for the policy decision emergence.

The decision emergence arises in the sub systemic environment of the policy decision process, and no policy theory deals with this unknown environment. Also, the concept of policy decision emergence is not developed so far and there are no theoretical directions to formulate the fuzzy inference rules. This development depends on each case and experimentation is necessary to refine the rules depending on each situation. Therefore, the development of the fuzzy rules is based on the expertise in the field and the judgment.

IF <PRB_Intensity = low>

AND <PRB_Extent = low>

AND <PRB_Urgency = low>,

THEN <Problem = low>.

Table 6.2 Fuzzy inference rules of FIS Problem (PRB) using AND operator

IF									THEN		
PRB_Intensity (input)			PRB_Extent (input)			PRB_Urgency (input)			FIS-PRB (output)		
L[a]	M[a]	H[a]	L	M	H	L	M	H	L	M	H
✓			✓			✓			✓		
✓			✓				✓		✓		
✓			✓					✓			✓
✓				✓		✓			✓		
✓				✓			✓			✓	
✓				✓				✓			✓
	✓		✓			✓			✓		
	✓		✓				✓		✓		
	✓		✓					✓			✓
	✓			✓		✓				✓	
	✓			✓			✓			✓	
	✓			✓				✓			✓
		✓	✓			✓			✓		
		✓	✓				✓		✓		
		✓	✓					✓			✓
		✓		✓		✓				✓	
		✓		✓			✓			✓	
		✓		✓				✓			✓
✓					✓	✓				✓	
✓					✓		✓			✓	
✓					✓			✓			✓
	✓				✓	✓			✓		
	✓				✓		✓			✓	
	✓				✓			✓			✓
		✓			✓	✓				✓	
		✓			✓		✓			✓	
		✓			✓			✓			✓

[a]L: *Low*, M: *Medium*, H: *High*

Table 6.3 Fuzzy inference rules of FIS Solution (SOL) using AND operator

IF									THEN		
SOL_Suitability (input)			SOL_Acceptability (input)			SOL_Feasibility (input)			FIS-SOL (output)		
L[a]	M[a]	H[a]	L	M	H	L	M	H	L	M	H
			✓			✓			✓		
✓			✓				✓		✓		
✓			✓					✓			✓
✓				✓		✓			✓		
✓				✓			✓			✓	
✓				✓				✓			✓
	✓		✓			✓			✓		
	✓		✓				✓		✓		
	✓		✓					✓			✓
	✓			✓		✓				✓	
	✓			✓			✓			✓	
	✓			✓				✓			✓
		✓	✓			✓			✓		
		✓	✓				✓		✓		
		✓	✓					✓			✓
		✓		✓		✓				✓	
		✓		✓			✓			✓	
		✓		✓				✓			✓
✓					✓	✓				✓	
✓					✓		✓			✓	
✓					✓			✓			✓
	✓				✓	✓			✓		
	✓				✓		✓			✓	
	✓				✓			✓			✓
		✓			✓	✓				✓	
		✓			✓		✓			✓	
		✓			✓			✓			✓

[a]L: *Low*, M: *Medium*, H: *High*

The fuzzy inference rules in this research are specifically formulated for the case study following an analysis of empirical data and a test of a certain number of assumptions.

As explained in Sect. 5.6.3, we built a decision truth table for each FIS of PODESIM. Each truth table uses the logical binary operator AND, and includes inference rules that represent all or most plausible scenarios (Tables 6.2, 6.3 and 6.4).

For example, the first row of the decision truth table for the FIS Problem (Table 6.2) represents the following rule (in the form of code below): If Problem Intensity is low and Problem Extent is low and Problem Urgency is low, then the problem has a low scale.

Table 6.4 Fuzzy inference rules of FIS politics (POL) using AND operator

IF									THEN		
POL_Mood (input)			POL_Cohesion (input)			POL_Leeway (input)			FIS-POL (output)		
L[a]	M[a]	H[a]	L	M	H	L	M	H	L	M	H
✓			✓			✓			✓		
✓			✓				✓		✓		
✓			✓					✓			✓
✓				✓		✓			✓		
✓				✓			✓			✓	
✓				✓				✓			✓
	✓		✓			✓			✓		
	✓		✓				✓		✓		
	✓		✓					✓			✓
	✓			✓		✓				✓	
	✓			✓			✓			✓	
	✓			✓				✓			✓
		✓	✓			✓			✓		
		✓	✓				✓		✓		
		✓	✓					✓			✓
		✓		✓		✓				✓	
		✓		✓			✓			✓	
		✓		✓				✓			✓
✓					✓	✓				✓	
✓					✓		✓			✓	
✓					✓			✓			✓
	✓				✓	✓			✓		
	✓				✓		✓			✓	
	✓				✓			✓			✓
		✓			✓	✓				✓	
		✓			✓		✓			✓	
		✓			✓			✓			✓

[a]L: *Low*, M: *Medium*, H: *High*

For the three FISs of PODESIM, the number of rules of each FIS represents all the possible scenarios according to our assessment. Each FIS has three input factors with three possible values each (Low, Medium, and High), therefore, the number of all combinations is $3^{**}3 = 27$.

Finally, for the FIS Emergence, we also built a decision truth table with the operator AND. This choice is inspired by Kingdon's multiple streams theory since it contributes to the development of the PODESIM model. Indeed, the multiple streams framework stipulates that the three streams (problem, solution, and politics) must come together to attract decision-makers' attention for a public issue. If we adjust this statement to fit the PODESIM requirements, we can say that there must be a conjunction of a problem, a solution, and a political will, for a decision to emerge. This concept is introduced in this research project through the application of the operator AND to the FISs (Problem, Solution, Politics) to formulate the inference rules of the FIS Emergence as the outcome of the system's interactions.

Based on a decision truth table of the FIS Emergence, we represent graphically the fuzzy inference rules of this FIS in Fig. 6.4 as set in Fuzzy Logic Toolbox^{TM}.

1. If (PRB is Low) and (SOL is Low) and (POL is Low) then (EMERGENCE is WEAK) (1)
2. If (PRB is Low) and (SOL is Low) and (POL is Medium) then (EMERGENCE is WEAK) (1)
3. If (PRB is Low) and (SOL is Low) and (POL is High) then (EMERGENCE is MED) (1)
4. If (PRB is Low) and (SOL is Medium) and (POL is Low) then (EMERGENCE is WEAK) (1)
5. If (PRB is Low) and (SOL is Medium) and (POL is Medium) then (EMERGENCE is MED) (1)
6. If (PRB is Low) and (SOL is Medium) and (POL is High) then (EMERGENCE is MED) (1)
7. If (PRB is Medium) and (SOL is Low) and (POL is Low) then (EMERGENCE is WEAK) (1)
8. If (PRB is Medium) and (SOL is Low) and (POL is Medium) then (EMERGENCE is MED) (1)
9. If (PRB is Medium) and (SOL is Low) and (POL is High) then (EMERGENCE is HIGH) (1)
10. If (PRB is Medium) and (SOL is Medium) and (POL is Low) then (EMERGENCE is WEAK) (1)
11. If (PRB is Medium) and (SOL is Medium) and (POL is Medium) then (EMERGENCE is MED) (1)
12. If (PRB is Medium) and (SOL is Medium) and (POL is High) then (EMERGENCE is HIGH) (1)
13. If (PRB is High) and (SOL is Low) and (POL is Low) then (EMERGENCE is WEAK) (1)
14. If (PRB is High) and (SOL is Low) and (POL is Medium) then (EMERGENCE is MED) (1)
15. If (PRB is High) and (SOL is Low) and (POL is High) then (EMERGENCE is HIGH) (1)
16. If (PRB is High) and (SOL is Medium) and (POL is Low) then (EMERGENCE is MED) (1)
17. If (PRB is High) and (SOL is Medium) and (POL is Medium) then (EMERGENCE is HIGH) (1)
18. If (PRB is High) and (SOL is Medium) and (POL is High) then (EMERGENCE is URGENT) (1)
19. If (PRB is Low) and (SOL is High) and (POL is Low) then (EMERGENCE is WEAK) (1)
20. If (PRB is Low) and (SOL is High) and (POL is Medium) then (EMERGENCE is MED) (1)
21. If (PRB is Low) and (SOL is High) and (POL is High) then (EMERGENCE is MED) (1)
22. If (PRB is Low) and (SOL is High) and (POL is Low) then (EMERGENCE is MED) (1)
23. If (PRB is Medium) and (SOL is High) and (POL is Medium) then (EMERGENCE is HIGH) (1)
24. If (PRB is Medium) and (SOL is High) and (POL is High) then (EMERGENCE is URGENT) (1)
25. If (PRB is High) and (SOL is High) and (POL is Low) then (EMERGENCE is HIGH) (1)
26. If (PRB is High) and (SOL is High) and (POL is Medium) then (EMERGENCE is URGENT) (1)
27. If (PRB is High) and (SOL is High) and (POL is High) then (EMERGENCE is URGENT) (1)

Fig. 6.4 Fuzzy inference rules of FIS Emergence

6.3 Case Study—Model Validation and Results

Following an experimental stage that allows us to verify that the computational model PODESIM is running well and able to generate reliable results, we proceeded with an empirical case study to validate the model.

6.3.1 Empirical Case Choice

We chose the Cuban missile crisis (October 15–28, 1962) as an empirical case to validate the PODESIM model. This case represents several facets that contribute substantially to this research. Indeed, the significant documentary sources related to this crisis and the analyses carried out by several theoretical approaches provide substantial detailed information to carry out a deep investigation of this case. Besides that, the crisis lasted a short period that facilitates the gathering and processing of data without cluttering the experimental model.

Moreover, and most importantly, we have access to a declassified primary source concerning the crisis. This source provides the most reliable information about crisis management and constitutes high-quality data for the simulation. From this perspective, the simulation of the Cuban missile crisis is an appropriate choice to validate the PODESIM model and to assess its strengths and limitations.

From a policy analysis perspective, the results of the simulation reveal an additional explanatory layer that brings out new details about the levers of the decision emergence and the decision-making during the Cuban missile crisis. Indeed, Allison and Zelikow (1999) argue that the "missiles of October" offer a set of fascinating puzzles for any analyst (p. 77), which reinforces our choice. Dobbs (2008), on his part, argues that there is a need for further study of the Cuban missile crisis, which has not yet revealed all its secrets.

Finally, Hass (2001) argues that the declassification of important information regarding this crisis, including recordings of meetings of US officials, could lead to more accurate interpretations and provide lessons from studying this crisis, which has been one of the most dangerous periods in recent history. Apart from these considerations, since the case study deals with a foreign policy case, it is relevant to cite Hudson (2005) who argues that explanatory variables from all levels of analysis, from the most micro the most macro, are of interest to the analyst to the extent that they affect the decision-making process. As a result, insights from many intellectual disciplines [...] will be useful to foreign policy analysts in their efforts to explain foreign policy decision making (p. 2).

6.3.2 Data Collection

The nature of the empirical case and available information necessitates qualitative research for data collection based on the requirements of the model. For this purpose, the PODESIM model constitutes a guide for data collection. Indeed, the model is intended for a specific objective in this research. Therefore, the model represents a reference and a basis to determine the data required for the simulation. We have, therefore, listed the most relevant works concerning the crisis and the decision-making on the American side, and we have concentrated our efforts on the factors of the PODESIM model that constitute the levers of the decision emergence within the American administration.

We based the data collection on a first-hand information source, the transcripts of the meetings of the Executive Committee (ExCom) of the National Security Council (May & Zelikow, 1997).[2] The transcripts are the authentic sound recordings of the White House secret meetings of the Crisis cell (ExCom) set up by President Kennedy.

Kennedy has secretly recorded the meetings without informing the ExCom members. Therefore, the transcripts of the recording provide factual and reliable information of considerable importance. It is in this primary source that we could find the facts reported by the major actors during the crisis. We did not consider any later descriptions or interpretations of those facts.

However, we completed and refined the data collection with a selection of complementary works considered as primary sources because they are reported by actors and direct witnesses of the crisis. These works add significant chronological details about the management of the crisis by the US Government.

The following references are used as complementary sources for data collection:

- Brugioni, D. A. (1991). Eyeball to eyeball. The Inside Story of the Cuban Missile Crisis. Random House, New York.
- Dobbs, M. (2008). One Minute to Midnight. Alfred A. Knopf. New York.
- Kennedy, R. F. (1968). Thirteen Days. The Cuban Missile Crisis. October 1962. Macmillan.
- Sorensen, T. C. (1965). Kennedy: The Classic Biography. Harper and Row, Publishers (Chap. XXIV).
- Sorensen, T. C. (2009). Minute by Minute: The Role of Intelligence in the Cuban Missile Crisis. International Spy Museum. Washington D.C.

We conducted an inductive thematic analysis based on the statistical frequency of words and sentences of the documentary sources, primarily of the ExCom meetings transcripts.

The objective of this analysis is to find links between qualitative textual content and the sub systemic factors of PODESIM. From this analysis, we were able to attribute, for each day of the crisis, a scale value to each factor of the model, at the lowest level ① represented in Fig. 6.1. The values are defined by the linguistic variables: low, medium, and high.

[2] May and Zelikow (1997).

Table 6.5 Factor coding method

Day	FIS	Factor	Fuzzy value	Documentary references
Day X	Problem	Intensity	Low	…
			Medium	…
			High	Example from ExCom meeting transcript: « […] the Soviet Union, for the past several days, has taken steps to bring its military forces into a complete state of readiness » (May & Zelikow, 1997, p. 348)
		Extent	Low	…
			Medium	…
			High	…
		Urgency	Low	…
			Medium	…
			High	Example: « I must inform you, however, that this is a matter of great urgency […] » (Kennedy, 1968, p. 157)
	Solution	…	…	…

The analysis of documentary sources allows us to build, for each day of the crisis, a tree structure representing medium-level variables (FISs) and their respective low-level factors. Then, using the NVivo® software, we compiled the textual occurrences of each factor to attribute a fuzzy value described by a linguistic variable that can be used for the simulation once linked to a value of a membership function.

Table 6.5 shows an example of data collection for a day X of the crisis and the FIS Problem.

The statistical compilation of collected data gave the following fuzzy values of the input factors of each FIS for each day of the crisis (Table 6.6).

Table 6.6 Fuzzy values of the factors of FISs (PRB, SOL, and POL) (Cuban missile crisis, 15–28 October 1962)

Date (October 1962)	FIS problem (PRB)									FIS solution (SOL)									FIS politics (POL)								
	Intensity			Extent			Urgency			Suitability			Acceptability			Feasibility			Mood			Cohesion			Leeway		
	L	M	H	L	M	H	L	M	H	L	M	H	L	M	H	L	M	H	L	M	H	L	M	H	L	M	H
15			✓	✓				✓						✓					✓			✓					
16	X	X		✓				✓		✓			X	X		✓			✓			✓			✓		
17			✓			✓			✓		✓		✓			✓			✓						✓		
18			✓			✓			✓	✓			X	X				✓	✓			X	X		✓		
19			✓			✓			✓	✓				✓				✓		✓		✓				✓	
20		X	X			✓			✓			✓	✓					✓	✓				✓				✓
21			✓			✓			✓			✓		✓				✓		✓			✓				✓
22			✓			✓			✓		✓			✓				✓			✓			✓			✓
23			✓			✓			✓			✓			✓			✓	✓					✓		✓	
24			✓			✓			✓				✓				✓		✓			✓			✓		
25		X	X		✓				✓			✓		X	X				✓				✓				✓
26			✓			✓			✓		✓			✓			X	X		✓			✓				✓
27			✓			✓			✓	✓				✓				✓		✓		✓				✓	
28	X	X			✓		✓			X	X			✓					✓			✓				✓	

L, M, H: Low, Medium, High
X Approximate equal Ratings

6.3.3 Simulation and Results

The simulation consisted of assigning to each factor various values within the range of values defined by the membership functions (Low, Medium, High) as defined in this chapter. For example, if a factor is considered low (0–25%), we simulate with different values within this interval to ensure that the simulation covers all the values in the interval.

This method gave a result in a range between a minimum and a maximum value of the simulated variable (FIS). These values represent a range of possible results and allow us to determine the average value of the output variable (FIS) that we used for graphical representations.

Practically, we proceeded as follows:

For each day of the crisis, considered as a step or "time step" in the simulation jargon, we carried out the simulations by entering, as inputs to the model, different step values of each factor according to the range of its membership function. The purpose of this approach is to test the impact of different values of each factor on the FIS to which it belongs.

For example, to test the influence of the factor "Suitability", in the range designated by the fuzzy scale "Medium", on the FIS Solution, we performed several iterations with a range of values (or steps) of the factor "Suitability" between 0.26 (26%) and 0.5 (50%), as determined for the scale "Medium".

For each step, the results given for each FIS (Problem, Solution, Politics) are converted to inputs of the highest level to perform the simulation of the FIS Emergence. This last step generates a numerical value of the FIS Emergence as the final output of PODESIM simulation.

Since PODESIM is a "prototype", we carried out simulations with several configurations using two types of membership functions for all factors and FISs: trapezoidal and Gaussian. For each configuration and simulation step, we also applied two defuzzification methods: the center of gravity (COG) method and the mean of maximums (MOM) method. These two methods provide slightly different outputs, which allows us to compare and evaluate the impact of each defuzzification method on the result.

The simulation provides significant information on the usefulness of each type of membership function and each defuzzification method.

The following figures illustrate a sample of the simulation using the Rule Viewer module of the Fuzzy Logic Toolbox^{TM}. The Rule Viewer module allows us to observe the simulation of each FIS and the impact of the input values in real-time.

The sample in Fig. 6.5 shows the values and membership functions of each input FIS (PRB, SOL, and POL) as well as the value of the output FIS EMERGENCE, after aggregation of the inference rules and defuzzification.

The simulation results with the configurations described above are represented in the following tables.

Configuration 1 Simulation using Trapezoidal membership functions and defuzzification methods COG and MOM (Table 6.7).

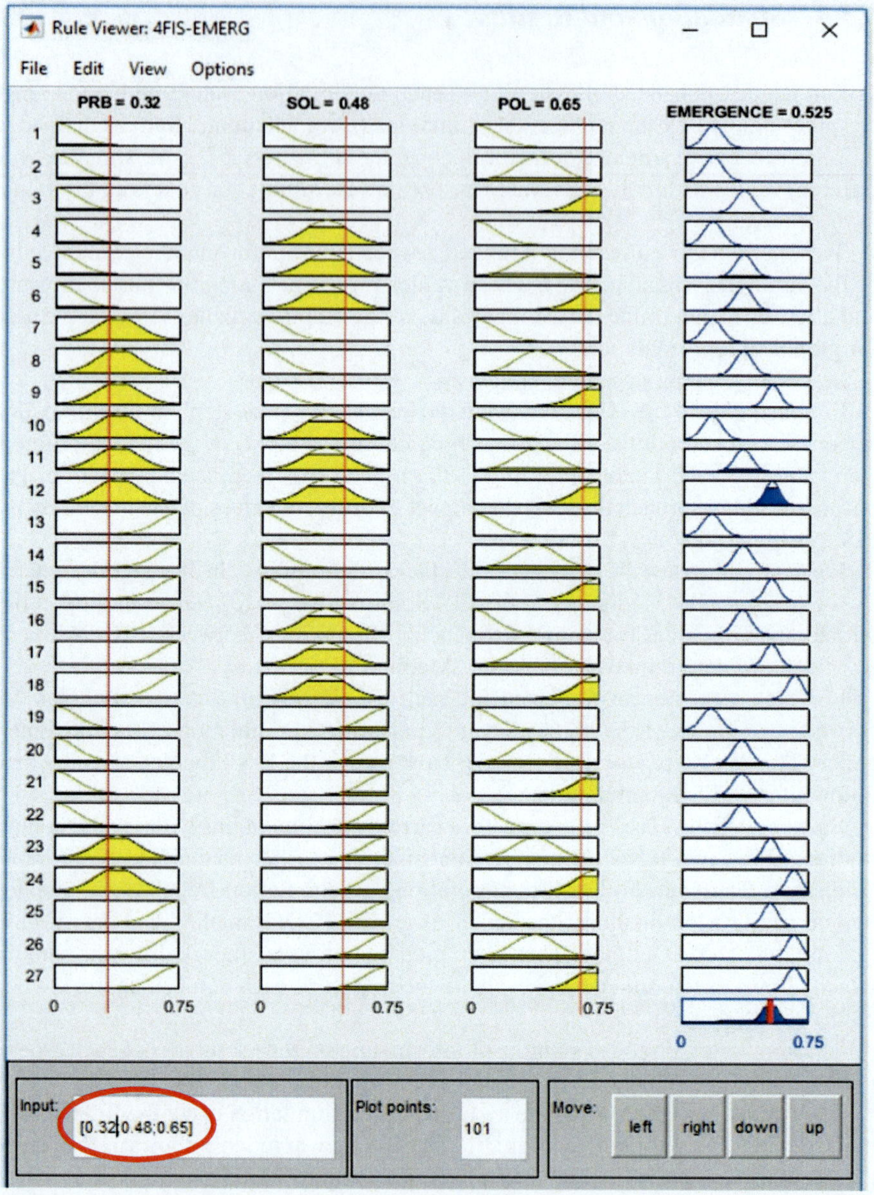

Fig. 6.5 Example of simulation of the FIS EMERGENCE

Table 6.7 Defuzzified output values of FISs of PODESIM

Date	FIS-PRB			FIS-SOL			FIS-POL			FIS-EMERGENCE	
	PRB (Min)	PRB (Max)	PRB (Med)	SOL (Min)	SOL (Max)	SOL (Med)	POL (Min)	POL (Max)	POL (Med)	EMERG (COG)	EMERG (MOM)
15	0.092	0.103	0.098	0.038	0.093	0.091	0.088	0.105	0.097	0.162	0.162
16	0.180	0.318	0.249	0.130	0.318	0.249	0.180	0.318	0.249	0.297	0.323
17	0.432	0.666	0.549	0.130	0.318	0.249	0.180	0.318	0.249	0.433	0.476
18	0.647	0.668	0.658	0.647	0.652	0.650	0.318	0.318	0.318	0.665	0.665
19	0.647	0.668	0.658	0.557	0.647	0.602	0.103	0.200	0.152	0.460	0.481
20	0.666	0.668	0.667	0.430	0.655	0.543	0.318	0.647	0.483	0.592	0.628
21	0.666	0.668	0.667	0.655	0.655	0.655	0.318	0.647	0.483	0.665	0.665
22	0.666	0.668	0.667	0.663	0.668	0.666	0.570	0.668	0.619	0.666	0.665
23	0.666	0.668	0.667	0.647	0.668	0.658	0.375	0.375	0.375	0.665	0.665
24	0.570	0.652	0.611	0.317	0.647	0.482	0.180	0.373	0.277	0.498	0.581
25	0.377	0.668	0.523	0.657	0.668	0.663	0.570	0.647	0.609	0.631	0.648
26	0.432	0.662	0.547	0.375	0.375	0.375	0.550	0.647	0.599	0.538	0.602
27	0.432	0.668	0.550	0.375	0.375	0.375	0.103	0.318	0.211	0.341	0.430
28	0.263	0.318	0.291	0.663	0.668	0.666	0.180	0.180	0.180	0.482	0.500

Table 6.8 Defuzzified numerical values of model variables

Date	FIS-PRB			FIS-SOL			FIS-POL			FIS-EMERGENCE	
	PRB (Min)	PRB (Max)	PRB (Med)	SOL (Min)	SOL (Max)	SOL (Med)	POL (Min)	POL (Max)	POL (Med)	EMERG (COG)	EMERG (MOM)
15	0.261	0.339	0.300	0.187	0.328	0.258	0.187	0.328	0.258	0.310	0.352
16	0.187	0.355	0.271	0.187	0.305	0.246	0.187	0.328	0.258	0.308	0.349
17	0.424	0.626	0.525	0.375	0.375	0.375	0.187	0.341	0.264	0.361	0.349
18	0.424	0.626	0.525	0.448	0.551	0.500	0.328	0.345	0.337	0.425	0.349
19	0.424	0.626	0.525	0.570	0.589	0.580	0.261	0.339	0.300	0.459	0.525
20	0.587	0.638	0.613	0.538	0.565	0.552	0.470	0.598	0.534	0.520	0.521
21	0.424	0.626	0.525	0.563	0.602	0.583	0.520	0.624	0.572	0.512	0.660
22	0.424	0.626	0.525	0.561	0.602	0.582	0.415	0.635	0.525	0.489	0.525
23	0.424	0.626	0.525	0.422	0.626	0.524	0.375	0.386	0.381	0.447	0.349
24	0.424	0.626	0.525	0.210	0.366	0.288	0.261	0.345	0.303	0.376	0.349
25	0.606	0.636	0.621	0.210	0.339	0.275	0.485	0.598	0.542	0.520	0.525
26	0.424	0.626	0.525	0.338	0.596	0.467	0.607	0.634	0.621	0.512	0.525
27	0.424	0.626	0.525	0.451	0.595	0.523	0.305	0.339	0.322	0.434	0.349
28	0.305	0.365	0.335	0.489	0.624	0.557	0.328	0.339	0.334	0.408	0.352

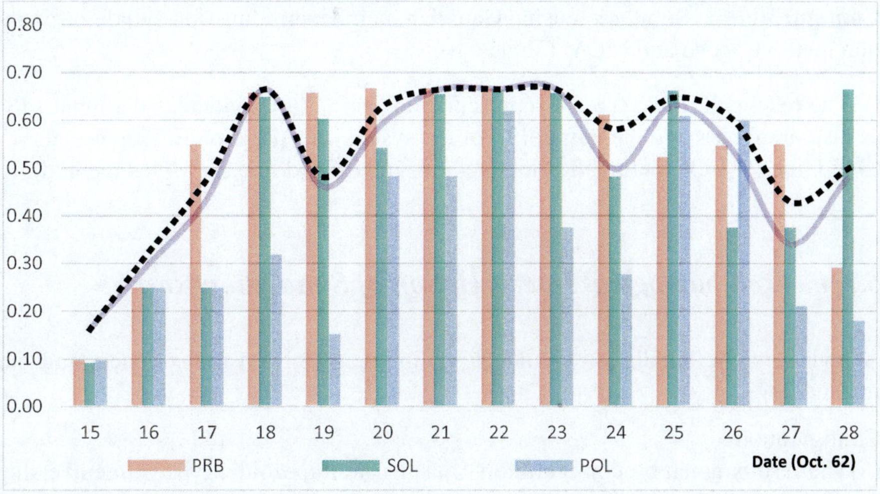

Fig. 6.6 Decision emergence and subsystems using Trapezoidal membership functions (Solid lines represents COG defuzzification method and dashed line represents MOM defuzzification method)

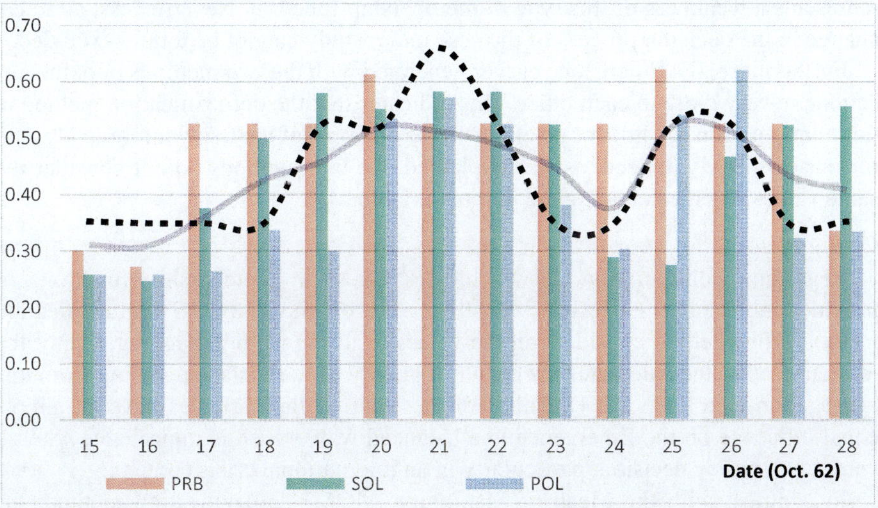

Fig. 6.7 Decision emergence and subsystems using Gaussian membership functions (Solid line represents COG defuzzification method and dashed line represents MOM defuzzification method)

These results are graphically represented in the Fig. 6.6 that includes the input FISs as histogram bars and the output FIS of the system as a decision Emergence curves (Solid line for COG defuzzification method and dashed line for MOM method).

Configuration 2 Simulation using Gaussian membership function and defuzzification methods COG and MOM (Table 6.8).

The results of Table 6.8 are represented in Fig. 6.7 that includes the input FISs as histogram bars and the output FIS of the system as a decision Emergence curves (Solid line for COG defuzzification method, and dashed line for MOM method).

6.3.4 Methodological Interpretation of Simulation Results

A review of the simulation results raises important observations concerning our methodology.

Configuration 1

The results given by configuration 1, using the trapezoidal type of membership functions, show that the Emergence curve displays a steep jump at the beginning and a constant phase in the center. This aspect can be explained by the fact that trapezoidal membership functions express linear behavior. Therefore, they are likely to go from a minimum to a maximum without considering the transition in the process, which represents a weakness of this type of membership function. Nevertheless, a radical change in the decision process of the case under study cannot be totally excluded.

Furthermore, the Emergence curves generated with the two methods of defuzzification are very close to each other. This indicates that the defuzzification method in this configuration has little or no effect on the outcome of the fuzzy inference system for our case study. Indeed, as was explained, the two methods do not consider the same values when calculating the output variable.

Configuration 2

Regarding configuration 2, which uses Gaussian membership functions, it provides more balanced results with the center of gravity (COG) defuzzification method. This method considers all the values resulting from the aggregation of the inference rules and calculates the center of gravity of the surface resulting from this aggregation (see Sect. 5.6.4). This indicates that all the variables have an almost equal influence on the Emergence in a balanced way, which is improbable when it comes to a policy decision, particularly in an international crisis context.

In contrast, with the Mean of Maximum (MOM) defuzzification method, the graphical representation displays a substantial variation of the Emergence.

This phenomenon is due to the very nature of the method, which has a discriminating power when calculating the output value of the fuzzy inference system. Indeed, it only considers the values generated by the aggregation of the inference rules that correspond to the maximums of the membership functions. This excludes the low values of the system.

This method seems more appropriate in a crisis context because it provides more realistic results. Indeed, in this context, only the most significant factors are generally of greater importance to the decision-makers because they cannot have all the information needed to assess all options and evaluate the impact of all factors in the decision process. Therefore, the discriminatory power of the MOM method makes this method more consistent with the case studied and more convenient to generate reliable results of the Emergence.

We conclude that configuration 2 using Gaussian membership functions combined with the Mean of Maximum (MOM) defuzzification method is an appropriate configuration for our empirical case. Moreover, Gaussian membership functions are, in this case, more adequate to capture fuzzy circumstances. This type of function seems to better represent complex non-linear systems using fuzzy information.

In the following chapter, we analyze the results obtained with this configuration.

References

Allison, G. T., & Zelikow, P. (1999). *Essence of decision: Explaining the Cuban missile crisis* (2nd Ed.). Longman: Addison-Wesley Educational Publishers Inc.

Bouchon-Meunier, B., Dotoli, M., & Maione, B. (2007). On the choice of membership functions in a Mamdani-type fuzzy controller.

Dobbs, M. (2008). *One minute to midnight: Kennedy, Khrushchev, and Castro on the brink of nuclear war*. New, York: Vintage Books.

Haas, M. L. (2001). Prospect theory and the Cuban missile crisis. *International Studies Quarterly, 45*(2), 241–270.

Hudson, V. M. (2005). Foreign policy analysis: Actor-specific theory and the ground of international relations. *Foreign policy analysis* (Vol.1) 1–30. Malden and Oxford: Blackwell Publishing.

May, E. R., & Zelikow, P. D. (1997a). *The Kennedy tapes: Inside the White House during the Cuban missile crisis.* Cambridge, MA: Belknap Press.

May, E. R. & Zelikow, P. D. (Eds.). (1997). The Kennedy tapes: Inside the White House during the Cuban missile crisis. Cambridge, Massachussets, and London, England: The Belknap Press of Harvard University Press.

Chapter 7
Analysis of Results

If everything on Earth were rational, nothing would happen...
Fiodor Dostoïevski.

The analysis of results is based on our statement in Sect. 6.3.4 regarding the methodological interpretation of results. The simulation results considered as reliable are obtained with a configuration using Gaussian membership functions and the mean of maximum (MOM) defuzzification method. This configuration better reflects the fuzzy reality of the policy decision emergence as a complex system in this research.

In this chapter, our objective is not to make a conventional analysis of the Cuban missile crisis using a linear estimating approach, but rather to examine the results provided by the simulation in the light of the facts and political events that affect the decision process during the crisis and to identify the decision emergence levers on day by day basis.

Our goal is primarily to examine, for each day of the crisis, the state of the system represented by the model PODESIM, and the decision emergence that results from the daily state of the system. More precisely, we want to identify the level of the decision emergence, and the scale of the variables of the system, i.e. the three FISs (Problem, Solution, Politics). Secondly, we aim to identify the factors that constitute the real levers of the decision emergence and to validate whether the results given by simulation are corroborated by historical facts and academic works about the Cuban missile crisis.

To undertake this analysis in an unbiased way, we chose a different set of academic works that deal with the analysis of the Cuban missile crisis than the references used for data collection.

Indeed, the primary source references report what the actors and the witnesses of the crisis said and did in the middle of the crisis, principally during the ExCom meetings. These references allowed us to qualify the fuzzy factors of the system for each day of the crisis. Therefore, they are not considered for analyzing the results of the simulation carried out with that "measurement" of the factors.

Besides, since the emergence reflects the decision circumstances likely to lead to decision-making, our analysis does not focus on a specific decision but on the state of the decision process and the decision-making environment. In other words, we

A. Guidara, *Policy Decision Modeling with Fuzzy Logic*, Studies in Fuzziness and Soft Computing 405, https://doi.org/10.1007/978-3-030-62628-0_7

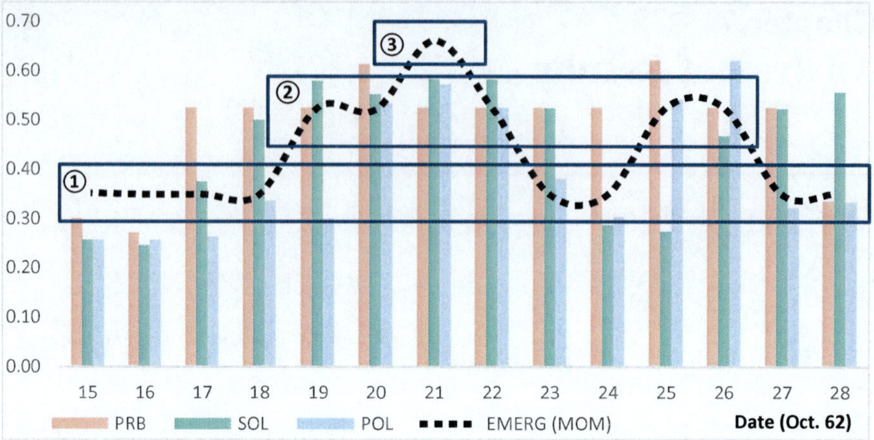

Fig. 7.1 Decision emergence and subsystems using gaussian membership function and mean of maximum (MOM) defuzzification method

examine the state of the system for each day of the crisis, to verify whether the level of the decision emergence indicates a decision made on that day and to identify the levers of the decision emergence and their impact on the decision-making.

As mentioned, this crisis is characterized by a variety of decision-making circumstances due to the diversity of possible responses of the US administration to the crisis. This diversity is also a result of the conflicting positions within the ExCom about the response to the crisis. This diversity has a tangible impact on decision emergence.

First and foremost, let us elaborate on the graphical representation of the results reproduced with more details in Fig. 7.1 to highlight the particularities of the findings.

The simulation of PODESIM using the appropriate configuration 2 displays three profiles representing three levels of the policy decision emergence. These levels correspond to three membership functions of fuzzy sets: medium, high, and urgent. From this graphical representation, we can draw the following statements about the policy decision emergence level during the crisis:

- Profile ①: this profile represents a Medium Emergence with values in the range of 30–40%. This profile concerns half of the crisis time (seven of fourteen days). It is the lowest level of decision emergence that represents a sub-decisional circumstances.
- Profile ②: illustrates High Emergence values (more than 40% and less than 60%) which corresponds to five non-successive days of the crisis. It represents a pre-decisional stage in this case.
- Profile ③: shows the highest level of Emergence for the duration of the crisis that occurs on October 21, 1962. This corresponds to the fuzzy range Urgent (more than 60%) and represents a decisional stage in which specific decision circumstances arise.

From these three profiles of decision emergence, a first observation arises about the relationship between system variables or FISs (Problem, Solution, and Policy) and the FIS representing the Emergence. Indeed, we can easily notice that the level of the FIS Emergence (curve) does not follow the state of the subsystems (the FISs) and their levels represented by histogram bars.

The results do not show any causal link or a correlation between the state of the FISs and the level of the Emergence. We believe that this link if it existed, could have been identified and investigated by analytical methods. This also demonstrates that our methodology and the choice of modeling and simulation are an appropriate approach to handle such a complex system. Indeed, the global behavior of such a system is not a sum of its components. This global behavior arises as a result of interactions that cannot be identified or analyzed with linear and deterministic frameworks.

Furthermore, the results also show that the Emergence has no low values in this case, although it is one of the four membership functions formulated for this FIS. We believe that this proves the reliability of the results in this case study, which deals with a serious international crisis, because decision-making is the most significant concern in such circumstances.

Similarly, we believe that the nature and gravity of the crisis imply that the decision emergence cannot reach very high values such as 75% or more despite the urgency of the situation. We believe that the decision circumstances in the context of a major international crisis cannot be favorable to a simple decision process and to a peaceful agreement among decision-makers. In such cases, there are so many factors, variables, and interactions that cannot be efficiently considered by the decision-makers. This aspect is also shown in the results, and we can see that the maximum level of the Emergence does not exceed 66%. Moreover, this value concerns only one day of the crisis.

These findings show the strength of PODESIM to generate consistent and realistic results for this case study.

In the following section, we analyze the daily situation of the system represented by PODESIM and the related events, and we examine the characteristics of the components of the model that explain the level of the decision emergence.

We use equally the terms variable or FIS to describe the subsystems of the model.

7.1 Day-by-Day Analysis

October 15, 1962

On the morning of the first day of the crisis, the American intelligence services confirmed the presence of Soviet weapons in Cuba, following an analysis of aerial pictures taken the day before by American U-2 planes. President Kennedy is informed.

Figure 7.1 shows that the decision emergence has a medium (lowest) level, and the three variables (FISs) of the system (Problem, Solution, and Politics) are relatively low. The Problem variable is even closer to the low range despite the discovery of Soviet missiles in Cuba. In fact, Soviet military activities in Cuba are not a surprise to the American administration. As early as the beginning of October, the US Government was considering preparations for a military operation against Cuba (Fursenko & Naftali, 2014). However, the situation and the nature of the crisis have not yet been precisely determined, and this is reflected in the state of the system and expressed by the level of the decision emergence.

It is also possible that the US administration still believed that the Soviet weapons being installed in Cuba are defensive, as proclaimed by the Soviet authorities. Indeed, Stern (2005) states that on September 11, 1962, the Soviet news agency TASS publicly insists that all weapons for Cuba were defensive and boasted that Soviet nuclear missiles were so powerful that it was unnecessary to look for sites outside the U.S.S.R. (p. 20).

These details are also reflected by the state of the system. The fact that the decision emergence level is not yet significant compared to other days of the crisis implies that the decisional environment does not yet have precise alternatives. This also means that the system is even not yet in a pre-decisional state.

An examination of the sub systemic factors indicates indeed that almost all these factors are low or non-existent, except the Intensity (of the problem) factor, which has a high value (Table 6.6).

This observation also confirms that the level of decision emergence that represents the overall behavior of the system does not depend on the individual state of the factors because it is a result of various dynamic interactions within this complex system.

October 16, 1962

It is on this date that a committee of the National Security Council is constituted and begins its discussions and deliberations. This committee is a crisis unit referred to by the acronym (ExCom[1] for *Executive Committee*).

Our results show that the levels of system variables (FISs) are relatively low. This is because the U.S. administration is not yet able to determine with certainty the magnitude of the problem. Indeed, the ExCom is still unsure of the nature of the threat and does not know what the Soviets are building in Cuba (Stern, 2005). Moreover, President Kennedy still does not understand why the Soviets have placed missiles in Cuba. Allison and Zelikow (1999) point out that at that time, which they consider to be the first day of the crisis, at least four separate times [...], Kennedy wondered aloud why the Soviets had done it. "Well," he shrugged, "it's a goddamn mystery to me." (p. 81). It wasn't until the ExCom meeting on the afternoon of October 16,

[1]For a list of ExCom members and their roles, see: Allison and Zelikow (1999). *Essence of Decision. Explaining the Cuban Missile Crisis.* Second edition, p. 326, 327. Longman. A more complete review is provided by Sheldon Stern (2005). *The Week the World Stood Still. Inside the Secret Cuban Missile Crisis.* Stanford University Press, Stanford, California, p. 29–35.

1962, that Secretary Douglas Dillon addressed to President Kennedy: «What if they carry a nuclear weapon?» (Stern, 2005, p. 47).

But these developments do not result in a significant change in the decision-making environment, as the situation remains vague and uncertain. Khrushchev informs US Ambassador Kohler that the presence of Soviet weapons in Cuba is not significant and stresses the defensive nature of Soviet military installations in Cuba (White, 1996). Finally, although the ExCom discussions focused on how to respond to the installation of Soviet missiles in Cuba, at the end of the day, a course of action was still far from clear [...] (Stern, 2005, p. 52).

As for Allison (1971) who argues that two major propositions emerged from these discussions: an air attack followed by a naval invasion or blockade, with the danger of subsequent military action. But these are only proposals pending the identification of the precise nature of the threat.

These clarifications are reflected in the results of the PODESIM simulation. We can see that decision emergence has not progressed from the day before. The system is still not in a sub-decisional phase, because the nature of the threat and the response to the situation have not been precisely determined.

Furthermore, regarding the uncertainty and ambiguity of the events at that time, White (1996) evokes a curious paradox in President Kennedy's position. On one side, Kennedy does not believe that the installation of Soviet missiles in Cuba would change the balance of power between the two countries, nor does he believe that Khrushchev would use these missiles against the United States. On the other side, he remains convinced that a military response to the Soviet challenge in Cuba is necessary. White concludes that perhaps the most plausible answer is that his eagerness for a military response flowed primarily from the need he felt to convey a sense of decisiveness, especially from the standpoint of domestic politics (p. 124).

These reflections suggest that President Kennedy, probably motivated by domestic political imperatives, is in no hurry to decide on Soviet missiles before analyzing all aspects of the crisis and calculating the pros and cons of the decision to be taken. We believe that these details are also reflected in the state of the system by the level of the decision emergence that remains below a level that could generate decision-making circumstances.

October 17, 1962

The results of the simulation show a system state characterized by a jump in the variable Problem. Indeed, American intelligence discovered two new Soviet facilities and confirmed the existence of medium-range ballistic missile sites. The variable Solution also exhibits an increase in its level. But, despite these changes, the level of decision emergence remains constant compared to the two previous days. Our analysis shows that these results are corroborated by events and that they converge with other analyses of the Cuban missile crisis.

Within the ExCom, no consensus has yet been reached on the appropriate response. Indeed, during the first days of the crisis, the analysis and evaluation of alternatives led to the establishment of a list of six major categories of action (Allison and Zelikow, 1999). However, this partial progress in the ExCom's deliberations is not reflected in a progression of decision emergence, nor in a progression of the variable Politics, which still has a low value.

However, these developments are reflected by an increase in the variable Solution, and this is probably not the only reason that has influenced this variable.

As a result of further deliberations that day, President Kennedy selected three alternatives that he felt were viable. These alternatives are an air attack limited to the missile sites, a general attack, and finally an invasion of the island of Cuba. However, he expressed his preference for a general air attack while advocating the possibility of restricting this action to a limited attack if necessary (White, 1996, p. 119). But these deliberations still do not have an impact on the decision emergence level.

Moreover, although the level of the variable Solution increases significantly, the factors of this variable have low values, except the factor Suitability (see Table 6.6). This is probably because since October 16, 1962, according to White (1996), JFK continued in the evening ExCom session to endorse the military options [...] [he] showed no interest in avoiding an immediate military confrontation by establishing a blockade around the island [...] (p. 120). The non-military proposals put forward by members of the ExCom were, in Kennedy's view, of no interest and not worthy of consideration (White, 1996), at least until October 17, 1962.

It should be mentioned that the United States Congress authorized President Kennedy, as early as October 3, 1962, to take the necessary steps to prevent the installation of Soviet military bases in Cuba, even if this required the use of force (White, 1996, p. 140). However, it is the Secretary of Defense, Robert McNamara, who is advocating the blockade as a response to the crisis and he is not alone. According to White (1996), Robert Kennedy emerged after October 16, as the most vigorous proponent of the blockade and impassioned critic of the air strike (p. 136). He thinks the military attack would damage the reputation of the United States.

In conclusion, the level of decision emergence given by the simulation indicates that it does not yet generate decision-making circumstances, despite a significant increase in the variables Problem and Solution.

October 18, 1962

The state of the system indicates a stationary medium-level of decision emergence, with even a slight decrease that keeps the decision circumstances at the sub-decisional stage. The results show that the variable Solution displays a strong increase and the variable Politics shows a small increase compared to the previous days. As for the variable Problem that keeps the same level as the previous day.

The increase in the variable Politics seems to be due to a slight improvement of the factor Cohesion. This aspect is raised by Allison and Zelikow (1999), who mention a near-consensus developed among Kennedy's advisers (p. 117) on that date. But Cohesion is the only factor of the variable Politics that shows a small increase according to our data, and this converges with other analyses.

In fact, White (1996) argues that on the morning of October 18, 1962, President Kennedy still believed that a general strike on Cuba was his best option. When his advisers talked about the merits of the blockade in that morning's ExCom meeting, he was skeptical (p. 89). However, according to Stern (2005), it was only late on the night of October 18, 1962, that President Kennedy stated that a consensus on a blockade began to emerge. Stern argues that JFK recalled, “opinions had obviously switched from the advantages of a first strike [...] to a blockade.” (p. 65). This is reflected in the progression of the variable Solution, probably driven by the only factor Feasibility.

However, Sorensen (1965) argues that the choice of the blockade was not a final decision as of October 18, 1962, and even less so the next morning, as Robert Kennedy continues to argue for a consensus in favor of a blockade among ExCom members. White (1996) claims in this regard that from October 17 to October 21, Robert Kennedy argument had remained constant: an American air strike on Cuba was reprehensible because it would be equivalent to the Japanese assault on Pearl Harbor (p. 149). White adds that by October 18, however, the president had withdrawn his backing for the air strike (p. 157).

Despite this evolution in President Kennedy position, which seems to strengthen the variable Solution due to a certain convergence of points of view within the ExCom, we can observe that decision emergence is still stationary and even slightly weaker than in the previous days. This is possibly due to the hesitancy that continues to affect the decision-making environment, that we raised above.

In addition, other interrogations arise. White (1996) argues that in ExCom meeting on the morning of 18 October, Acheson and his colleagues started to ponder the legal ramifications of the various policy options (p. 142).

It is the result of the analysis of the legal aspects that has greatly contributed to strengthening the blockade option. This option has proved to be the easiest and quickest solution to implement. This aspect is reflected in our examination of the factor Feasibility which contributes to the progression of the variable Solution.

The conclusion to be retained from October 18, 1962, is that the decision emergence level does not evolve, despite the strong progression of the variables Solution and Policy, raised above. This progression is due to two factors: the factor Cohesion and the factor Feasibility but without any impact on the level of decision emergence.

These observations demonstrate once again that decision emergence has a complex system behavior characterized by the absence of a causal link between its components and its global behavior. The significant increase in the three variables of the system and some factors has no impact on the decision emergence. This emergence shows a stationary state since the beginning of the crisis, and there was no decision made until October 18, 1962, or a serious alternative considered.

These conclusions support our decision heuristics (represented by the inference rules) and demonstrate that they are suitable for this case study.

We can therefore conclude that during the first four days of the crisis, the state of decision emergence remains constant and that it does not reach a significant level that generates decision-making circumstances. This level is represented by the subdecisional Profile ①, which describes an absence of a decision emergence, despite a progression of variables and factors.

October 19, 1962

On that date, the level of decision emergence starts to progress. It displays a level in profile 2. The variables (FISs) of the system have different states. The variable Problem is at the same high-level as the previous two days, whereas the variable Solution shows great progression, at a higher level than the variable Problem. Finally, variable Politics is slightly decreasing.

According to Stern (2005), the consensus around the blockade, established late in the night of October 18, 1962, seems to set the stage for the next day and provide President Kennedy with greater assurance in the continuation of meetings and the decision-making process. The author argues that Kennedy ignored [General Maxwell] Taylor and began to speak immediately in a clear effort to demonstrate that the commander-in-chief was in charge (p. 67). However, this event does not seem to increase the variable Politics, but rather in a slight decrease, which we can link to the weakness of the factor Cohesion (see Table 6.6) on October 19, 1962.

From a political perspective, President Kennedy must demonstrate his firmness in managing this crisis because of the election campaign. Allison and Zelikow (1999) argue that the Republicans Senatorial and Congressional Campaign Committee announced that Cuba would be the "dominant issue of the 1962 campaign." (p. 330). But what about the solution being considered?

Stern (2005) points out that President Kennedy, before leaving for Ohio and Illinois for the election campaign, urged Robert Kennedy and Sorensen to forge a consensus around the blockade during his absence (p. 71). He also adds that McNamara endorsed planning for air strikes but continued to support the blockade as a measured first step (p. 71). These details indicate that President Kennedy has already chosen the blockade while continuing to spare some ExCom members who have a rather belligerent attitude. May and Zelikow (2002) argue that [...] in his October 19 meeting with the Joint Chiefs of Staff in which, alone, he takes on the combined weight of their arguments and has apparently gone far toward making up his mind (p. 440). As for White (1996) who claims that it was on October 19, 1962, that a consensus concerning the blockade emerged thanks to the efforts of Robert Kennedy and McNamara (p. 173).

These statements allow us to conclude that the blockade option chosen by President Kennedy as a solution to the crisis is the most important element underlying decision emergence. This element affects both variables, Politics, and Solution. The evolution of the situation shows that President Kennedy is playing a central role and that he is granting himself greater leeway. He favors the non-military option, unlike many members of the ExCom. Stern (2005) also states that if the ExCom decisions had been made by majority vote then war, very likely nuclear war, would almost certainly have been the result (p. 5). This aspect highlights one of the characteristics of American politics where the President's position is very important. These details seem to impact the progression of the factors Leeway and Mood, but without resulting in a progression of the variable Politics, which shows the absence of a causal link between the low-level factors and the middle-level variable, the FIS Politics. This variable too behaves as a complex system.

The variable Solution sharply raises, certainly because of the preliminary choice made by President Kennedy. Its progression is driven by the factor Feasibility which seems to be the decision emergence lever on this date.

We can then conclude that the most significant levers of the progression of the decision emergence are in this order, the factor Feasibility, followed by the factors Acceptability, Leeway, and Mood. This emergence has a high level and illustrates a pre-decisional stage that we have represented by high profile 2 in as illustrated in Fig. 7.1.

October 20, 1962

The results indicate a level of decision emergence like the day before, but with a slight decrease. However, the variables (FISs) display different patterns compared to the previous day. The variables Problem and Politics are on the rise. On the other side, the variable Solution shows a slight decrease.

Part of the explanation can be found in Stern's (2005) statement that during his meeting with the ExCom on October 20, 1962, President Kennedy was backing away from the blockade option. He also argues that the defense chief acknowledged that the blockade might create "political trouble" at home, [...] and a blockade was less likely to provoke a Soviet response "leading to general war." (p. 72). For his part, Robert Kennedy supports the combination of the military option and the blockade (Stern, 2005). However, the members of the ExCom remain divided on which option to adopt.

We can see that despite the hesitations, the blockade option remains the preferred alternative for many ExCom members, and this is probably what explains the progression of the variable Politics that shows a huge increase supported by the factor Cohesion, which is on the rise.

Indeed, Stern (2005) states that Kennedy authorized the blockade and suggested that "we inform the Turks and the Italians that they should not fire the strategic missiles they have even if attacked." (p. 73).

Stern adds that President Kennedy had, however, agreed to continue preparations for an invasion, stating that after the ExCom meeting, JFK chatted on the second-floor balcony with RFK and Sorensen. "We are very, very close to war," he conceded bleakly (p. 74). This attitude seems to be reflected in the progression of the variable Politics, as the President has chosen the blockade and we can link this progression to the high level of the factor Leeway. But Kennedy's choice remains hypothetical. It must be decided and well-defined.

Furthermore, Allison and Zelikow (1999) assert that, on that day, two different options for the blockade were presented to President Kennedy and they specify that the blockade was only chosen, however, after the option was sharpened into the blockade and ultimatum approach on October 20-21 (p. 120). Nevertheless, they consider the date of 20 October 1962 to be the key date of the decision to go with the blockade (p. 117).

Despite these statements, the results indicate a slight decrease in the variable Solution. This is probably due to persistent disagreements among ExCom members. According to Stern (2005), Robert Fitzgerald Kennedy shifted ground again, arguing that a combination of a blockade and air strikes "was very attractive to him." [...] Suddenly, most remaining participants took sides: Rusk endorsed the blockade; McCone, Dillon, and Gilpatric essentially agreed. McNamara warned that air strikes would kill thousands of Russians and Cubans and "the U.S. would lose control of the situation." General Taylor still dissented (p. 73).

All these details suggest that despite the progress made in the discussions, the U.S. administration is still hesitant about the final decision.

The disagreements within the ExCom are also reflected in the factor Mood, down on the previous day, but with no negative impact on the variable Politics, which is rising sharply. Another element that may explain the increase of the variable politics is the role played by the American Ambassador to the United Nations, Adlai Stevenson, who elaborated a memorandum to Kennedy and the ExCom. According to White (1996), in this memorandum, he described a "Political Program" to be announced by the president and developed by himself in the UN Security Council at the same time as the imposition of the blockade. Such an approach would convince the international community, which might otherwise view the quarantine as needlessly provocative, that the United States was intent on reaching a peaceful settlement (p. 174).

Stevenson calls for a Soviet-American dialogue to negotiate a permanent solution to the problem and a Security Council resolution to deploy an observer mission to Cuba, Italy, and Turkey. According to White (1996), he also calls for a simultaneous withdrawal of Soviet missiles and military personnel from Cuba and the US bases in Guantánamo Bay, Turkey, and Italy. White claims that his arguments on the need to blockade Cuba and on the utility of working through international organizations elicited general agreement from ExCom officials (p. 175). These elements suggest that Stevenson's arguments have been convincing and have strengthened the variable Politics with an increase that we can link to the factor Cohesion, which shows some progression.

However, Stevenson's proposals for American concessions are strongly criticized by his colleagues, who are rather furious at the idea of making so many concessions to the Soviets. According to White (1996), even President Kennedy sharply rejected the thought of surrendering our [Cuban] base. [...] He felt that such action would convey to the world that we had been frightened into abandoning our position (p. 175).

We believe that these contradictions among officials have weakened the variable Solution, following a decline of the factor Acceptability.

White (1996) also adds that [Stevenson] was able to convince neither Kennedy nor any other officials of the need for a quick diplomatic solution to the crisis before it escalated into military conflict[30] (p. 175). Many of those who could have supported him that day were absent and Stevenson had to make his case in ExCom alone, and, consequently, his arguments did not acquire the sort of legitimacy they would have enjoyed had they been embraced by others (p. 176).

According to White (1996), This weakened Stevenson's position and cast doubt on his ability to adequately represent his country at the United Nations during this crucial period. White also adds that Robert Kennedy even told Schlesinger, who accompanied Stevenson to United Nations headquarters, we're counting on you to watch things in New York, [...] That fellow is ready to give everything away (p. 176).

All these elements seem to further delay formal decision-making, despite the high level of decision emergence on October 20, 1962.

To end this very eventful day, Sorensen (1969) argues that it was only at the end of the day that Kennedy seriously opted for the blockade and according to Stern (2005), he suggests to inform Italy and Turkey and ask them not to retaliate in case of Soviet aggression (p. 73). Nevertheless, he continues to support the preparations for a possible invasion of Cuba.

Apart from the factors in the variable politics that have remained high since October 17, 1962, we can conclude that the levers of the high decision emergence of October 20 are the following factors: Feasibility, Suitability, and Leeway. This emergence shows a pre-decisional profile 2.

October 21, 1962

The decision emergence is at its peak, with a level defined as urgent. It displays a decisional stage with a Profile 3. The state of the system indicates a decrease in the variable Problem and an increase in the variables Solution and Politics. It demonstrates in the first place that the highest level of decision emergence does not necessarily require the highest values of the three variables (FISs) of the system, and it highlights, once again, the complex nature of the system in which there are no direct causal links or correlations between its components.

On another level, few changes affect the factors of the three variables. Except for the factors Acceptability and Mood, which both change from low to medium level, all other factors remain constant compared to the day before. This observation further shows the absence of causal links or correlations within the system. How does a little change, which barely affects only two of the factors at the sub systemic level of the system, result in a maximum level of decision emergence?

The paradigmatic answer lies in the evolutionary dynamics of this complex system, where a small change at the micro-level is likely to have a significant impact at the macro-level. This impact cannot be predicted, and it depends on the overall dynamics of the entire system. These dynamics result from the interactions of the components to varying degrees. The simulation results demonstrate that it is on that date that the decision is made because of the highest level of decision emergence.

These results are corroborated by specific events. Indeed, Stern (2005) argues that in the afternoon of Saturday 21 October 1962, during the ExCom meeting in the Oval Office, the deliberations had entered a new phase. The quarantine had been chosen [...] Rusk suggested that it would be useful to call the blockade a "quarantine", because "it avoids comparison with the Berlin blockade." (p. 75).

May and Zelikow (2002) provide more details about this choice. They claim that the blockade was only chosen, however, after the option was sharpened into the blockade and ultimatum approach on October 20-21 [...] Kennedy had started leaning toward the blockade option as early as October 18, but on October 22 he explained to his National Security Council, "from the beginning, the idea of a quick strike was very tempting and I really didn't give up on that until yesterday morning [...]" (p. 154).

President Kennedy was about to announce to the nation the next day the discovery of the missiles and his intention to impose a blockade that would take effect on October 24, 1962. May and Zelikow (2002) also point out that the American administration chose the next day to officially inform its allies, members of Congress, and ambassadors of friendly and neutral countries. Finally, according to Stern (2005), it was also on this day that the President also sent personal representatives to brief the leaders of Britain, France, and West Germany (p. 76).

All these details confirm that the final decision was made on October 21, 1962, and corresponds to the maximum value of decision emergence given by simulation. The fact that the results are consistent with the events of that date reinforces the PODESIM capacity to simulate the decision emergence and justifies the choice of the type of membership functions and the defuzzification method selected for this case study. Moreover, the results also validate the fuzzy inference rules formulated for this case study.

However, despite this conclusive result, some elements remain unexplained since the factors of the system have not changed much from the day before. We cannot attribute the result to the only two factors that have progressed from the day before. An examination of the variables of the system could guide us towards identifying the levers of the decision emergence.

The results show that the variable Problem decreased, although data shows that the factors of this variable remained constant. The explanation could be linked to the evolution of this variable over time due to its complex nature, but this evolution cannot be explained at this point. We can only state that the overall behavior of this variable cannot be deduced from the individual state of its components.

As for the variable Solution, it shows an increase compared to the previous day. This increase is clearly due to the slight increase in the factor Acceptability.

This is also the case for the variable Politics that shows an increase due to a slight rise in the factor Mood. As for any complex system, one should not exclude that a slight change in a single component at the micro-level may have a significant impact on the overall behavior of a system. However, these clarifications do not allow us to deduce that the factors Acceptability and Mood alone constitute levers for decision emergence.

We can therefore conclude that the decision emergence is due to a combination of several factors that cannot be precisely determined. But, compared to the day before, it is the factors Acceptability and Mood that have a definite impact on the progression of the variables Solution and Politics, and consequently on the evolution of the system that has generated this level of decision emergence. This emergence shows a decisional stage that corresponds to favorable decision-making circumstances.

October 22, 1962

The results display a sharp decline in the decision emergence level. We can deduce that this significant decline is because the decision has already been made and it is about to be announced, but that the events and dynamics of the decision-making environment are still able to generate new decision-making circumstances. The state of the system indicates that the variables Solution and Problem do not change from the previous day, but that the variable Politics is slightly decreasing.

This date represents a frenetic day for Kennedy (White, 1996). The President is about to announce, in his address to the nation, the discovery of Soviet missiles in Cuba. Two hours before his address to the nation, President Kennedy receives a special delegation of congressmen and some senators as he seeks bipartisan support for his decision (Fursenko and Naftali, 2014). Kennedy's cautious handling of the crisis and his refusal to declare war against Cuba and the Soviets annoyed Democratic Senator Richard Russell, who advocated war, which disoriented Kennedy and his Defense Secretary Robert McNamara (Fursenko and Naftali, 2014).

Did this event contribute to the weakening of the variable politics? It would be tempting to think so, but two of the factors of this variable are rather on the rise. The factors Mood and Cohesion are increasing from medium to high level, but with a negative impact on the variable Politics. We think that the decrease of the variable Politics is related to the factor Leeway.

Nevertheless, there are also other events worthy of interest. Before his address to the nation, Kennedy stressed to the ExCom members the importance of staying united and told them «Everyone should sing one song.» (White, 1996 p. 185). Despite Kennedy's attitude to bring ExCom members together around a common goal, the variable Politics still decreased.

According to Sorensen (1969), Kennedy's decision to opt for a blockade and his rejection of the military option did not please all the members of the ExCom, some of whom preferred to fight with the Soviets. Sorensen argues that [...] the Joint Chiefs of Staff had been dangerously inflexible in their insistence on an all-out military attack as the only course the Soviets would understand [...] (p. 192). This detail could explain in part the decline of the variable Politics, especially since the President is required to constantly justify his choices. White (1996), for example, claims that President Kennedy, in his conversations with ExCom members, states that the danger that a failure on his part to respond would damage the United States position throughout Latin America because it would then appear as though "the Soviets were increasing their world position while ours was decreasing." (p. 185).

Several other elements could help to clarify this situation.

The American administration was not sure how the Soviets would react to the blockade. According to Stern (2005), President Kennedy told ExCom members on the afternoon of 22 October 1962 that Khrushchev will *not* take this without a response, maybe in Berlin or maybe here [...] (p. 80).

It is also on this date that the crisis becomes international and puts the world in front of this explosive situation. This could explain why the decision emergence remains at a high level but without reaching an urgent scale.

Moreover, Sorensen (1969) states that Kennedy refused to issue an ultimatum, to close any doors, or to insist upon any deadlines, noting only that continued work on the missile sites would "justify" (not necessarily insure) further U.S. action (p. 189). On the same date, President Kennedy publicly called for an immediate meeting of the Organization of American States (OAS) and an urgent meeting of the United Nations Security Council. However, the American administration cannot yet know what the international reaction will be, especially that of the Soviet authorities. It is probably for this reason that President Kennedy is asking the members of the ExCom to reflect and decide on what will happen next, once the blockade is in place. According to White (1996), Kennedy asks, among other things, that if the missile development in Cuba continues [after the establishment of the quarantine], what is our next course of action? (p. 186). However, according to Stern (2005), Kennedy also clearly said "we can't invade Cuba," because it would take days to assemble the necessary forces (p. 87).

To conclude, we can affirm that it is the evolution of the system over time and the interactions of several factors of the three variables that determined the level of decision emergence. We can retain that the factors Mood and Cohesion registered an increase and the factor Convenience shows a decrease. But these changes alone cannot explain the evolution of the system and the level of decision emergence.

We presume that after the decision was made on 21 October 1962, the various events suggest that the situation within the ExCom and on the international stage created new conditions that seem to bring the system to a different phase. In this new phase, a new context and a different mindset start to prevail and do not allow to clearly circumscribe the new contours of the decision-making environment. This new situation goes beyond the US administration and this case study. We would be tempted to consider certain factors as levers for the decision emergence on that day, but their determination seems difficult and risky, and this aspect represents a limitation of PODESIM in this case study.

October 23, 1962

It is on this date that the level of the decision emergence decreases significantly as well as the level of the variables Solution and Politics.

Yet, it is on this date that the Organization of American States (OAS), of which Cuba is not a member, meets and approves the measures taken by the US government. The Council of this Organization adopts a resolution, based on articles 6 and 8 of the OAS treaty, that supports the decisions of the US administration and demands the dismantling and withdrawal of missiles and other weapons from Cuba.

Regarding our model, despite an increase in certain factors, such as the Suitability and Acceptability, this results in a minor decrease in the levels of variable Solution, and the decision emergence.

This may be explained by the fact that the Soviet defense minister, Rodion Malinovsky, announced on October 23, 1962, that a decision had been made to increase the preparation and vigilance of Soviet troops. In addition, the Commander-in-Chief of the Warsaw Pact, Marshal Andrei A. Grechko, called for a meeting of officers from several Eastern European countries. White (1996) claims that Grechko instructed them to heighten the military readiness of their forces (p. 190). Meanwhile, several other Soviet ships are heading for Cuba. White (1996) reports that the Soviet military attaché in Washington informed reporters on the evening of October 23, 1962, that these ships were ordered to ignore the blockade and continue their way to Cuba. These elements had certainly weakened the variable Solution since the sequence of events suggests that the response chosen by the American administration may not be the best one, for it seems to lead to an escalation.

On another plan, in a letter to Kennedy dated October 23, 1962, Khrushchev shows no sign of turning back. White (1996) claims that [he] told Kennedy that the establishment of a blockade was "aggressive" and a "threat to peace." [...] Offering no concessions, Khrushchev instead called on Kennedy to revoke his recent decisions, which, he indicated, could have "catastrophic consequences for world peace (p. 190, 191). These details are likely to explain the sharp decline of the variable Politics, caused by a sharp decrease in the factors Mood and Leeway, and this results in a sharp drop in the decision emergence. Only the factor Cohesion remains at its high level (see Table 6.6).

In conclusion, some serious events of October 23, 1962, and the internationalization of the crisis seem to greatly disrupt the decision-making environment and shaken the decision emergence process. The situation on this date seems rather confused and does not facilitate the rise of decision emergence circumstances. This situation provoked a significant decline in the decision emergence to profile 1, a sub-decisional stage.

We can also conclude that the decision emergence no longer depends on the US national environment and the ExCom alone, but it depends on several new factors, some of them are related to the role of the international organizations involved in this crisis, whereas other factors depend on the Soviet Union government. This new situation reduces the US Administration's room for maneuver and it is represented by the decline of the factor Leeway.

Finally, it was on October 23, 1962, that the first meeting of the UN Security Council related to the crisis took place. A meeting that allowed the American ambassador, Adlai Stevenson, to defend the position of the United States. It was followed by the response of the Soviet ambassador to the United Nations as well as the Cuban representative. Under these circumstances, it is quite logical that the decision emergence is at a sub-decisional stage.

October 24, 1962

Decision emergence is still at the same level after its fall on the previous day and it still shows a sub-decisional profile 1, while the variable Solution is falling drastically as well as the variable Politics. The state of the factors is consistent with the evolution of these variables since they too are all falling, several factors have low values. But this development does not imply a causal link between the factors and the variables and consequently the decision emergence.

As in the previous two days, the variable Problem remains high, probably because Khrushchev showed no signs of accommodation in his 24 October message to Kennedy, [...] according to White (1996). Khrushchev [also] declared that the Soviet-Cuban relations were none of America's business, charged Kennedy with trying to intimidate Moscow into submission, and claimed that the blockade was illegal (p. 200). This is supported by Allison and Zelikow (1999) who also argue that Khrushchev's message to Kennedy on October 24 was defiant. They claim that in his message, Khrushchev said he would tell Soviet Captains to ignore the American quarantine (p. 124). Khrushchev also denies the right and authority of the OAS to adopt a resolution that allows the United States to enforce the blockade. He expresses his rejection of the resolution which he considers a violation of the freedom of navigation in territorial waters and aggression that pushes humanity towards nuclear war.

On the ground, new pictures taken by American spy planes indicate that work continues unabated at missile sites and McNamara informs the ExCom that a Soviet submarine is approaching the demarcation line of the blockade (Preston, 2001, p. 127). The US administration is no longer certain how to manage the implementation of the blockade if Soviet ships refuse to comply with it. In the meantime, President Kennedy wants to avoid a military confrontation with ships that would ignore the blockade, despite the opposition of some ExCom members. McNamara, for his part, states that it is probably necessary to prepare for a confrontation (Preston, 2001).

These elements clearly explain the uncertainty and the instability of the situation, which has become partly uncontrollable, with a real impact on the variables of the system. These aspects are the consequence of the decline of the factors of variable Politics.

In relation with the decline of the variable Politics, it is important to mention that the Secretary-General of the United Nations U Thant sent a letter on October 24, 1962, to the American and the Soviet leaders suggesting that the United States suspend the quarantine and the Soviet Union cease weapons shipments to Cuba for two or three weeks in order to provide a block of time for a settlement to be reached (White, 1996, p. 206). Secretary U Thant's suggestion was enthusiastically received by Khrushchev, but Kennedy suggested that Ambassador Stevenson discuss the proposal further with the UN Secretary-General.

This is shown by the sharp drop of the variables Solution and Politics, which keeps the decision emergence at the lowest sub-decisional level. Again, during this stage, the levers of decision emergence are not significant to generate decision-making circumstances, and besides that, the US administration must consider several new external factors since the internationalization of the crisis.

October 25, 1962

This date marks a jump of the decision emergence to the high level, as well as an increase of the variables Problem and Politics, but only a very small decrease of the variable Solution.

White (1996) argues that U Thant dashed off new messages to the American and Soviet leaders on 25 October. He asked Khrushchev to keep his ships away from the quarantine line for "a limited time only," and urged Kennedy to "do everything possible to avoid direct confrontation with Soviet ships in the next few days" (p. 207). It is this aspect that reinforces the variable Politics, driven particularly by the factor Leeway, which is progressing at a high level after a drastic drop the day before. Moreover, the same day, Kennedy expressed his acceptance of U Thant's proposal, if Khrushchev also accepts it. This situation seems to support the progression of the decision emergence, which now shows a profile 2, a pre-decisional stage.

Regarding the variable Solution, the slight decrease of this variable can be explained by the fact that Kennedy continued to avoid any action that might provoke a hostile Soviet response (White, 1996, p. 200). But he eventually recognized that the pressure from the blockade might not be sufficient to persuade Khrushchev to back down (p. 201).

Both Kennedy and many ExCom members thought that the Soviets would retreat in return for the promise to withdraw US missiles from Turkey, but events have shown otherwise. This situation brings the US administration back to the starting point, and the ExCom must now elaborate on other alternatives.

President Kennedy is now considering further measures against the Soviet leadership, and in doing so, he is aligning himself with the position of some ExCom members who, from the beginning of the crisis, advocated an aggressive response and refused to make any concessions to the Soviets. It is probably this attitude that partly explains the progression of the variable Politics, from a low level to a medium level due to the factor Cohesion.

Kennedy ends up taking a hard line by giving nothing to the Soviets. White (1996) claims that in his message to Khrushchev on 25 October, Kennedy, [...], simply defended his position, condemned his adversary's, and introduced no terms of settlement to the crisis (p. 201). However, no new solution seems to emerge as of October 25, 1962, to ease the crisis. This aspect is represented in the result by the significant increase of the variable Solution.

Nevertheless, the absence of a new solution is not the only aspect, because the variable Problem is also increasing. Indeed, the Soviet ship Bucharest, transporting oil to Cuba, was intercepted by the US Navy, but immediately released because it was not carrying weapons. But this act shows that the United States is ready to use force to enforce the blockade. Moreover, quoting McNamara, May, and Zelikow (2002) state that referring to the Soviet missiles, he comments on October 25, "I never have thought we'd get them out of Cuba without the application of substantial force." (p. 442).

These authors also argue that the turning point of the crisis may have been October 25, the day that Khrushchev decided that he would withdraw the missiles on terms that would abandon his most important original goals for the deployment (p. 443).

For this date, it is difficult to draw reliable conclusions about the levers of decision emergence. The decision-making environment remains ambiguous and dependent on domestic and international factors. However, we can presume that the factors Convenience, Cohesion, and Leeway have largely contributed to increasing the level of the decision emergence, at least as far as the US administration is concerned.

October 26, 1962

Decision emergence remains high, identical to the day before, and shows a predecisional profile 2. But the simulation results show a strong increase of the variables Politics and Solution and a decrease of the variable Problem.

Let's analyze what happened on this date. White (1996) claims that during the ExCom meetings of October 26, 1962, JFK continued to advocate the application of pressure on the Soviets while avoiding action that might evoke a hostile response from Khrushchev (p. 207). He ordered the arrest and inspection of the Maricela, the first vessel intercepted since the beginning of the blockade, to demonstrate the United States' determination to strengthen the "quarantine". However, Khrushchev ordered not to replicate (White, 1996). It is Khrushchev's reserve that might explain in part the decline of the variable Problem.

Nevertheless, the Marucla incident had repercussions within the ExCom, because during the discussions it was concluded that the blockade would not be enough to persuade the Soviets to withdraw the missiles from Cuba. Opinions diverged as to what should be done in order to achieve the objective set by the United States. The debate relates largely to military action against the missiles and a possible embargo, especially as work on the Soviet bases in Cuba continues with no interruption, according to information obtained the same day.

At the end of the discussions, it was decided that if the negotiations at the UN did not provide a quick settlement, "our choice would be to expand the blockade or remove the missiles by air attack." (White, 1996, p. 208). This seems to confirm our findings. Indeed, the high level of decision emergence can be explained by this decision regarding the next steps in the sequence of events.

It is also this situation that explains the sharp increase in the variables Solution and Politics. The variable Politics now seems to be supported by the increase of the factors Mood and Leeway (see Table 6.6) and these elements are also confirmed by White's statements above.

Regarding the decrease of the variable Problem, it occurs when the factors of this variable are all at a high level, with an increase of the factor Extent. Another event could explain this decline. It concerns the Soviet Government who loosens its grip and proposes a solution to end the crisis. Khrushchev decided to dismantle the launching bases in Cuba, in exchange for an American promise not to invade Cuba. Garthoff (quoted in Nathan, 1992) argues that Khrushchev continued for a few days to believe that the United States might accept at least the partial Soviet missile deployment already in Cuba. But by October 26, it had become clear that the United States was determined (p. 47).

On the evening of October 26, Khrushchev sent a message to Kennedy in which he proposed the reciprocal withdrawal of Soviet missiles from Cuba and American missiles from Turkey, and called for a commitment by the United States not to invade Cuba. We believe that this event influenced all the variables in the system and is well represented by the results of the simulation. It is mainly the variable Solution that shows a significant increase, despite the average level of its factors and even a decrease in the factor Suitability.

Finally, the last element that explains the state of the decision emerging, which displays a pre-decisional profile 2, is undoubtedly the following: on the evening of 26 October 1962, the ExCom met to discuss the response to give to Khrushchev's message and it was decided to postpone the decision until the following day. President Kennedy and some of his advisers were reluctant to accept the American commitment not to invade Cuba, as requested by the Soviets.

October 27, 1962

Our results show a sharp decline in the decision emergence level, while only the variable Politics undergoes a sharp decline. According to White (1996), it is "Black Saturday" in Washington. Work on missile sites in Cuba continues and some sites are considered fully operational. CIA reports reveal a rapid mobilization of the Cuban army and a memorandum indicates that the Soviet mission in New York is destroying important documents, which would indicate a possibility of imminent armed conflict. These events keep the variable Problem at the same level as the previous day. This variable does not increase, probably because its factors remain unchanged.

On another plan, the news caused confusion within the US administration. According to Preston (2001), this confusion is accentuated by Khrushchev's position. Indeed, the Soviet leader, pushed by the military staff, publicly declares that he is asking for the dismantling of the American military bases in Turkey to put an end to the crisis and he informed the American administration by sending a letter to Kennedy, while the ExCom is preparing to respond to Khrushchev's letter sent the previous day, not without stormy exchanges and friction among its members.

The US administration was disturbed and disunited by the successive messages sent by the Soviets on 26 and 27 October 1962, which included different proposals (White, 1996; Allison and Zelikow, 1999; Preston, 2001; Stern, 2005).

All these events explain the state of the variable Politics, which undergoes a sharp decline, probably caused by the falling factors Cohesion and Leeway. These two factors may be behind the sharp decline of decision emergence, especially if we reflect on the day of October 27, 1962.

Indeed, on that day around noon, an American U-2 plane was shot down over Cuba and its pilot killed, leading to an internal crisis within the ExCom. President Kennedy, positioned against the will of the military, banned any military response unless the incident was repeated and involved more US planes.

On this date, there is also a slight increase in the variable Solution, according to our results. This increase can be explained by the fact that Kennedy and his advisers agreed on a secret disarmament of US military bases in Turkey (White, 1996). It is probably the factor Feasibility that explains this decision.

However, in response to Khrushchev's letter of the previous day, the U.S. administration informed the Soviets that it agreed to end the blockade and promised not to invade Cuba against the withdrawal of Soviet missiles, but without addressing the issue of U.S. missiles in Turkey.

Furthermore, no decision has been made regarding subsequent actions, as the U.S. administration is not yet sure of the Soviet reaction and some ExCom members are already advocating an invasion. White (1996) argues that notes taken by Lyndon Johnson express the sense that many ExCom officials had about the likelihood and imminence of an American assault on Cuba: "regarding the peace in the Caribbean - By strike no later than Mon a.m. Invasion." (p. 220). White (1996) adds that Kennedy himself was acutely aware of the feeling in his administration that the time for military action against Cuba was approaching.

At several points in the ExCom meetings on 27 October, he spoke of the likelihood that the United States would have to attack Cuba in the next few days (p. 221). But he points out that during the discussions, it became clear that Kennedy viewed military action against Cuba as a last resort. "I'm not convinced yet of the invasion" he told his advisers. Rather, he was interested in trying two other approaches before considering that alternative (p. 222).

Despite the opposition of many ExCom members, Kennedy would rather withdraw American missiles from Turkey in exchange for the withdrawal of Soviet missiles from Cuba which could start an armed conflict. White (1996) states in this regard that although JFK's comments in ExCom and his decision to dispatch Bobby Kennedy to Dobrynin showed that he was anxious on 27 October to avoid a military conflict (p. 226).

These elements show that the decision-making circumstances are far from ideal and that the US administration still does not know how the situation will evolve or how to respond to the events. Meanwhile, Allison and Zelikow (1999) note that McNamara called to active duty 24 troop-carrier squadrons on the Air Force reserve, approximately 14,000 men. Thus, the blockade was but the first step in a series of moves that threatened air strike or invasion (p. 123). This situation is shown in the results with a decreasing decision emergence to its lowest level. The emergence displays again a sub-decisional profile 1 which seems to characterizes adequately this day of the crisis.

It is only on the evening of 27 October 1962 that the situation became clearer. According to Garthoff (quoted in Nathan, 1992), Kennedy's proposal [...] to exchange American assurances against invasion of Cuba for Soviet withdrawal of its missiles, coupled with a virtual ultimatum, was thus promptly accepted (p. 48). This is what White (1996) describes as a final settlement of the crisis, originally proposed by Ambassador Stevenson on October 20, 1962.

White argues that by refining Stevenson's ideas, Kennedy might have produced a settlement acceptable to both Khrushchev and American public opinion, thereby securing an early resolution to the crisis (p. 182).

It is this outcome that seems to be reflected in the increase of the variable Solution, due to the high level of the factor Feasibility, according to our data.

October 28, 1962

For the last day of the crisis, at least officially, the system's variables continue to influence the decision emergence. According to May and Zelikow (2002), the crisis is not entirely over. The agreement between the two powers must be followed by concrete actions to end the crisis, which represents continuity in the decision process. Moreover, White (1996) claims that many American officials continued to feel that military action against Cuba was virtually unavoidable (p. 226), especially since a CIA report indicates that work on Soviet missile sites has not stopped. However, these events do not seem to influence the variable Problem, which decreases to its lowest level, probably caused by the low levels of the factors Intensity and Urgency.

Furthermore, the variable Solution displays a sharp increase that can be explained by the turn of events. Indeed, Radio Moscow broadcasted a message of Khrushchev, in which he declared that the Soviets accepted the proposed settlement of dismantling and withdrawing Soviet weapons in exchange for the American commitment not to invade Cuba. But he did not raise the issue concerning the American missiles installed in Turkey. May and Zelikow (2002) claim that with this announcement, the goal set by the US administration was achieved.

Yet, the variable Politics remains at the same low level as the day before, probably due to the low level of the factors Mood and Cohesion. Indeed, Khrushchev's message is received in Washington at 9 a.m. According to White (1996), although some ExCom members welcomed the news with relief, others expressed disappointment and anger at this development (p. 229). Therefore, the system remains in a sub-decisional state and does not result in a decision emergence. Indeed, no decision was made by the US administration on the last day of the crisis.

Table 7.1 represents a synthesis of results given by PODESIM simulation of the decision emergence for the duration of the Cuban missile crisis.

Table 7.1 Synthesis of results

Date	Decision emergence level, profile, and tendency	Significant levers of decision emergence
October 15–18	**Medium** (profile 1) ---	
October 19	**High** (profile 2) ↗	**Feasibility (SOL)** **Acceptability (SOL)** Leeway (POL) Mood (POL)
October 20	**High** (profile 2) ↘	**Feasibility (SOL)** **Suitability (SOL)** **Leeway (POL)**
October 21	**Urgent** (profile 3) ↗↗↗	**Acceptability (SOL)** **Mood (POL)**
October 22	**High** (profile 2) ↘↘	Mood (POL) Cohesion (POL)
October 23	**Medium** (profile 1) ----	
October 24	**Medium** (profile 1) ----	
October 25	**High** (profile 2) ↗	Convenience (SOL) Cohesion (POL) Leeway (POL)
October 26	**High** (profile 2) ↘	Mood (POL) Leeway (POL)
October 27–28	**Medium** (profile 1) ---	Feasibility (SOL)

7.2 Conclusion

The results given by PODESIM simulation are largely corroborated by the events during the Cuban missile crisis, and by the actions of the U.S. administration which constitutes the decision-making environment. However, like any simulation model, PODESIM does not provide comprehensive results to explain all aspects of the case study, because a model is a simplified representation and not an accurate reproduction of the real system.

For this reason, the results must be taken with caution, although they have largely explained the state of the decision process in this case study, and reveal the decision emergence levers at the sub systemic level, which is the main goal of this research project.

To summarize our findings related to decision emergence, we retain that the decision emergence during the Cuban missile crisis displays three levels that we explain in the following manner:

- Sub-decisional profile (1) illustrates a decision-making environment that does not meet the minimum conditions for a decision emergence, even if the system variables (Problem, Solution, Politics) have high values. Decision emergence of this profile has a medium-level value, but the lowest in this case study, that does not generate decision-making circumstances.
- Pre-decisional profile (2) that represents a high-level decision emergence. The circumstances and some aspects of the decision-making environment indicate that the system is getting closer to a decision emergence, but is missing triggers for that.
- Decision emergence profile (3) that generates a decision emergence that can lead to decision-making. The decision emergence is at the highest level and it illustrates urgent circumstances for decision-making.

These profiles do not depend on the individual state of the variables (FISs) or the factors of the system. We demonstrated that there is no causal relationship between the level of decision emergence and the state of the variables and the factors. Also, there is no causal relationship between the factors and the variables neither. The results confirm that the decision emergence dynamic is typical of a complex system. In such a system, the systemic evolution at the macro-level is a result of dynamic interactions between the components at the sub systemic or micro-level. In this case study, we confirmed that the global behavior of the system, which represents the FIS Emergence, cannot be predicted from the individual state of its components.

7.3 PODESIM Limitations and Further Development

Despite the good quality of the findings and the validation of the methodology developed in this research, the PODESIM model has some limitations because it is still an experimental "prototype" for modeling and simulation of policy decision emergence. Among the limitations, we can state the following:

Firstly, the simulation is carried out with one empirical case study so far to validate the model. This validation is a necessary step in any modeling and simulation approach. However, it represents a limitation because the model needs additional empirical case studies to be further validated and generalized and to acquire certain robustness.

Secondly, for this case study, the model showed a limitation concerning the dates of October 22 and 23, 1962, because the decision process goes beyond the simulation frame that includes variables and factors limited to the only American government. We mention the need to add new factors and variables related to the international environment, even if we believe that the international environment influenced the ExCom debates and indirectly the data we collected for the case study. Such new factors may have an impact on the results. For that, one needs to explore other sources that relate to the international environment, and chose new components according to parsimony rules that respect a compromise between the real world and the model requirements.

Adding factors and variables is a promising initiative to complete and generalize the model. Indeed, the modular structure of PODESIM, built with blocks of fuzzy inference systems, allows the integration of additional FIS and inference rules without affecting the flexibility and the versatility of the model. As a model of a complex system, PODESIM remains an open system and can be completed with new components at multiple levels.

However, the theoretical background of public policy analysis that inspired the model does not suggest a way to integrate new factors and variables to the model. We believe that adding new components to the model requires empirical case studies in order to identify new factors and variables. This aspect opens a new research topic.

Furthermore, for this case study, we proceeded with the simulation without weighting the factors or the variables. Further case studies are needed to discover techniques that would allow the factors and variables to be weighted depending on the situations. Weighting also may have an impact on the results. The development of weighting techniques in this field is a potential research subject.

Additionally, we proceeded with fuzzy inference rules formulated with the only logical operator AND as inspired by the streams' coupling hypothesis formulated by Kingdon. Despite the good quality of findings, additional case studies are essential to determine whether these rules are robust in similar situations. However, the determination of fuzzy inference rules should be based on theoretical concepts and appropriate methodologies or empirically proven techniques. At this stage of research, no theoretical or empirical basis allows us to determine the fuzzy inference rules for a policy decision process. This can only be based on judgment and expertise, and on a case by case basis.

Furthermore, the membership functions chosen in this case study are based on our knowledge in the field of public policy and foreign policy. We believe that more experimentation with robust data and reliable time series lead to building more precise and specific membership functions. This is also a research topic that needs further development.

Finally, it should be noted that the case study in this research is based on qualitative research which is not free from subjectivity, and it represents an inherent weakness as for any qualitative research.

From a public policy analysis perspective, since the model is based on multiple streams framework, it would be relevant to carry out a comparative study between results given by PODESIM and other analyses based on multiple streams theory. Such a comparison could identify links between the notion of decision emergence and Kingdon's coupling of streams for example, or a relationship between the level of decision emergence developed in this research and the streams' coupling force raised by Lemieux (2009).

Finally, although the inference rules defined in this case study are inspired by Kingdon's streams coupling rule, this coupling phenomenon cannot be deduced from the results given by PODESIM. We demonstrated that no causal link exists between the level of decision emergence and the state of the variables or streams. Our findings do not bring a new contribution to the notion of streams coupling, but they do demonstrate that our modeling and simulation approach clearly stands out from the multiple streams theory. In this research, Kingdon's approach only constitutes a theoretical bridge between public policy and complex system theory that allows us to build a fuzzy inference model.

References

Allison, G. T. (1971). *Essence of decision: Explaining the Cuban missile crisis*. Boston: Little, Brown.

Allison, G. T., & Zelikow, P. (1999). *Essence of Decision: Explaining the Cuban Missile Crisis* (2nd Ed.), Longman (Addison-Wesley Educational Publishers Inc).

Fursenko, A., & Naftali, T. (2014). *"One Hell of a Gamble": Khrushchev, Castro, and Kennedy, 1958–1964*. New York, London: W. W. Norton & Company.

Lemieux, V. (2009). *L'étude des politiques publiques: Les acteurs et leur pouvoir*. Québec: Les Presses de l'Université Laval.

May, E. R., & P. D. Zelikow, (Eds.). (2002). *The Kennedy Tapes: Inside the White House During the Cuban Missile Crisis*. W.W. Norton, Concise Edition.

Nathan, J. A. (Ed.). (1992). *The Cuban Missile Crisis Revisited*. Palgrave Macmillan.

Preston, T. (2001). *The President and His Inner Circle. Leadership style and the advisory process in foreign affairs*. New York: Columbia University Press.

Sorensen, T. C. (1965). Kennedy. Harper & Row, Publishers.

Sorensen, T. C. (1969). The Kennedy Legacy. Macmillan.

Stern, S. M. (2005). *The Week the World Stood Still. Inside the Secret Cuban Missile Crisis*. Stanford University Press.

White, M.J. (1996). *The cuban missile crisis*. UK: Palgrave Macmillan.

Chapter 8
Innovation and Contributions

To dare; that is the price of progress.
Victor Hugo.

In this chapter, we outline the innovations of this multidisciplinary research and its contributions to the fields of decision studies and public policy.

The significant advancement brought by this research consists in the integration of several methodologies and the genuine combination of several theories and techniques that have not been combined so far in the field of public policy. This integration led to the development of a decision diagnostic model, designated by the acronym PODESIM (Policy Decision Emergence Simulation Model). The development of this model is achieved through this integration as well as the identification and characterization of the sub systemic components of the policy decision process and constitutes a major innovation.

Conceptually, the PODESIM model establishes theoretical and methodological links between distinct approaches and fields to complete and enrich the understanding of the decision process, and to explore the dynamics of the sub systemic environment of this process.

PODESIM contributes to overcoming the separations between fields that belong to different scientific cultures and paradigms. It also goes beyond the limits of traditional approaches by capturing the ambiguous aspects of the policy decision-making environment. This environment is largely affected by the subjectivity of the decision-makers as raised by March and Olsen (1985) but it remained unexplored because of the absence of appropriate tools.

PODESIM integrates into the decision process new notions from the "artificial sciences", a set of concepts advocated by Simon (1969) and considered as pioneering work in the field of artificial intelligence in the social sciences. This research brings a real contribution to this field and it advances the conceptual and the computational features.

On another level, this multidisciplinary development constitutes scientific progress because it combines two established theories, complex systems theory, and multiple streams theory, as the first advancement of this research in the field of policy decision studies. Cairney and Jones (2015) argues that multiple streams

A. Guidara, *Policy Decision Modeling with Fuzzy Logic*, Studies in Fuzziness and Soft Computing 405, https://doi.org/10.1007/978-3-030-62628-0_8

approach has made important contributions to policy theory and to the empirical litterature, but that these contributions remain separate from each other and do not constitute a coherent and robust body of research. This research provides part of the response to this criticism by integrating concepts of multiple streams theory and empirical methodologies. This work proved that this integration is coherent and it brings new research topics for further development.

Additionally, this work pushes back the limits of existing approaches by transforming the narrative multiple streams theory into an open modular computational model. This accomplishment contributes to theoretical and empirical advancement concerning the characterization and operationalization of the policy decision process.

Another innovative aspect of this work consists in a transition from systemic and actor-centered decision analysis at the macro level, to a new approach based on sub systemic factors and their dynamics at the micro-level of the decision process. This new approach goes beyond the limits reached by conventional decision-making approaches. It develops new tools able to capture the sub systemic environment of the decision process while integrating its properties as a complex system. The complex properties of this sub systemic environment have been raised by several policy studies, but they remained so far unexplored and undetermined.

This work made big progress by grasping the complexity of the sub systemic level of the decision process for it has a great influence on the decision emergence and consequently on the decision-making and the policy cycle. These aspects are very important and one should grasp them with great interest. According to Koliba and Zia (2011), complexity has always been a part of everyday public management and policy practices [...]. The extent to which a policy and governance system is stable or experience instability should matter to those interested in public administration and policy (p. 2, 3). Also, Klijn and Snellen (2009) claim that the history of the field of public administration could be viewed as an ongoing attempt to search for concepts to grasp the complexity of day-to-day practices in policy-making and decision-making (p. 17).

These statements allow us to assert that our approach responds partially to a real need not yet fulfilled in the field of public policy. This approach, based in part on complexity theory, explores new aspects of the decision process and reveals facets and details unexplored before. Meek (2010) argues that understanding and explaining complex aspects of the decision process represent a high potential to improve our understanding of decision-making and policy development and implementation by the public administration.

Also, in our approach, we use modeling with fuzzy logic as a method to simulate the decision emergence that raises at the sub systemic level of the decision process and to identify the levers of the decision emergence. The use of modeling and simulation with fuzzy logic represents a methodological innovation in the field of policy decision studies that contributes to the understanding of policy decision dynamics.

This methodological innovation also constitutes a paradigm shift. Indeed, the development of PODESIM represents a transition from a static decision-making analysis using classical approaches, to a dynamic approach that explores the mazes of the policy decision process, and reveals the dynamics that generate decision-making circumstances.

This transition also affects the decision-making diagnostic method, as our approach represents a shift from the classic top-down analysis to a bottom-up simulation of elements and dynamics at the sub systemic environment of the decision process.

Understanding the micro-level of the decision process that has complex properties in a bottom-up approach using modeling and simulation also results in an epistemological innovation. Indeed, this transition is a paradigm shift from positivist methods to a postpositivist approach. In other words, it is a shift from a rational and deterministic analytical approach to a generative approach, which is the essence of complexity (Epstein, 2006). This new paradigm offers new ways of understanding and describing the world (Mitleton-Kelly, 2003; Mitchell, 2009).

Beyond the paradigmatic and methodological advancement, PODESIM and the simulation technique, in general, constitute an added value in the field of public policy decision studies. Simulation plays an important role in forecasting and planning various scenarios and options to optimize the decision-making and its impact on policies, hence, it increases the quality and success of policies. This applies to the whole policy cycle because all stages of the cycle are dynamic and include uncertainty and ambiguous aspects that are not controllable nor predictable. These aspects can be captured through modeling and simulation.

Simulation of several scenarios and alternatives can also target specific issues when it is supported by good quality data. This results in proactive policy development that addresses properly public issues and policy development instead of reactively coping with public dissatisfaction and complaints (Tsoukias, A., Montibeller, G., Lucertini, G., & Belton, 2013, p. 126). This suggests that future evolution is considered because the long-term consequences of policy-making imply the need to consider the range of possible futures, sometimes characterized by deep uncertainties and calling for the development of future scenarios (Tsoukias et al., 2013, p. 128).

PODESIM can simulate different options which leads to properly structure the issues and to formulate optimized solutions to these issues. Tsoukias et al., (2013) argue that a large part of the decision support activities occurring within a policy cycle are about understanding, formulating, and structuring 'problems' (p. 129). They also add that most of the policy cycle is about designing or constructing alternatives. Actually, most of "smart" policy making is about "innovative design" of "innovation policies" (Montibeller & Franco, 2011), [...]. (p. 129).

Finally, PODESIM and, more generally, modeling and simulation extend the range of possibilities in public policy. Indeed, simulation can be carried out in real-time when provided with consistent and contextualized flow of data. This results in rapid decision-making and in the development of appropriate policies that target specific public issues. The combination of data, modeling and simulation, public policy, and administration represents an invaluable advantage that results in a better diagnostic of the decision emergence and strong support for decision-making and policy cycle in general.

Informed decisions and policies lead to appropriate implementation of the policies by the public administration and help to address the specific issues that are required to develop the policies. Modeling and simulation can also play a useful and important role in using information and data, from a variety of sources, to provide decision-makers and practitioners with a solid and specific groundwork for decision-making and managerial practice that no other technique is currently able to offer.

In Fig. 8.1, we illustrate how PODESIM can be integrated into the public policy and administration environment. In this figure, we also illustrate the role that data from different sources can play in the decision process and the public policy and administration.

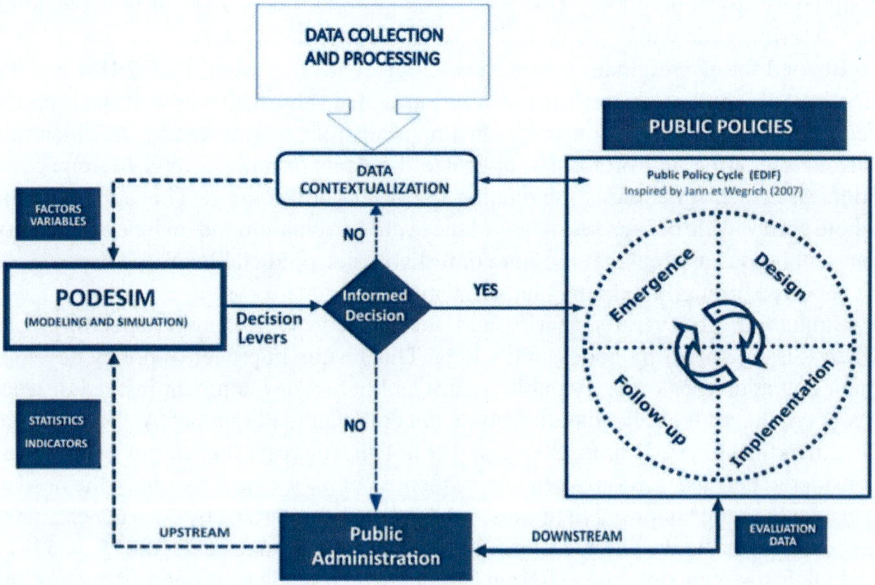

Fig. 8.1 PODESIM model in the policy and administration structure

If the decision process simulation is based on appropriate models and comprehensive data from reliable sources, it results in suitable decisions and data-informed policies. It improves the quality of public policies and strengthens their implementation and results. According to Janssen et al. (2015), simulation allows decision-makers to understand the essence of a policy, to identify opportunities for change, and to evaluate the effect of proposed changes in key performance indicators (Banks, 1998; Law & Kelton, 1991) (p. 7). These authors add that policy-making is confronted with increasing complexity and uncertainty of the outcomes which results in a need for developing policy models that can deal with this. To improve the validity of the models policy-makers are harvesting data to generate evidence (p. 2). PODESIM supports the managing of this complexity by using contextualized data from various sources to simulate the real world. It can therefore provide a more realistic and detailed picture of the decision-making environment to better capture what should constitute the "essence of the decision" as formulated by Allison (1971), through the identification of the decision emergence levers as a piece of evidence to justify decision-making and to legitimize policy development and implementation.

On another level, when implementing and evaluating policies, public organizations can use data collected to simulate and monitor results in real-time and adjust actions accordingly. This allows the public administration to be efficient, adaptive, and proactive with the support of data.[1] Public organizations can also contribute to a better governance through a dynamic based on data generation and sharing in an open public environment.[2]

In this perspective, PODESIM represents a concept and a fundamental basis for capturing not only the levers of decision emergence but also other different situations with specific arrangements and needs. Being an open, modular, and adaptable model, it can be developed further and modified for monitoring and forecasting when it is provided with appropriate and reliable data. Finally, it can also be used to carry out the evaluation of policies that could bring detailed and valuable information to decision-makers which results in specific and rapid responses to public issues because it allows linking evaluation results to the levers of decision emergence.

Consequently, it contributes to a more objective assessment of policies and more accurate measurement of their performance, as such assessments would be based on specific variables, precise criteria, and targeted objectives. Ultimately, this leads to making government intervention more efficient, and policy development and implementation more legitimate and transparent. Tsoukias et al. (2013) claim that as already introduced by Habermas (1981), legitimated policies are the ones which are appropriately explained, justified, supported and not sufficiently confuted (i.e. argued) (p. 128).

In other words, as a concept, the model can be adapted and used at all stages of the policy cycle. This concept is elaborated through a unique combination of public policy and complex systems concepts, fuzzy logic as an artificial intelligence method, and modeling and simulation using advanced computational techniques. This combination is a major innovation brought by this research in the field of decision studies in public policy. PODESIM offers a cutting-edge method in policy decision simulation, and it brings a unique contribution to emerging interdisciplinary fields, such as computational public policies (Solo, 2014).

---

[1]See on this subject: Mitchell and Mitchell (2016). *Adaptive Administration: Practice Strategies for Dealing with Constant Change in Public Administration and Policy* (ASPA Series in Public Administration and Public Policy). CRC Press.

[2]Concerning the needs and management of information resources and their impact on governance and public administration, see: Caron (2018). Chapter 26. La production documentaire dans les administrations publiques: enjeux et pistes de solution. In Michaud, N. (Dir.), *Secrets d'États? Les principes qui guident l'administration publique et ses enjeux contemporains.* 2e édition. Presses de l'Université du Québec.

PODESIM could also be grafted to the emerging field of Policy Analytics, a research area of crucial importance for the future of policy-making and an essential complement to classical policy analysis.[3]

In the public policy field, advancements using forefront methodologies and multidisciplinary integration remain limited while this field is at a crossroads where the gap between the past and the future of public policy studies is growing. Nowadays, this integration is necessary and it must be developed and implemented to meet the real needs of policy-making, digital transformation, and governance.

Zgurovski and Zaychenko (2016) argue that the use of fuzzy inference systems (FIS) [...] has allowed to solve many problems of decision-making under uncertainty, incompleteness, and qualitative information—forecasting, classification, cluster analysis, pattern recognition (p. 81). These assertions obviously support our methodology and encourage further use and improvement of PODESIM.

A multidisciplinary integration is not only essential to make the policy cycle more efficient, but it also generates new ideas and techniques that shed new light on the reality of dynamic public policies. Such integration also enhances the exchanges among different fields and constitutes a source of mutual enrichment.

In the multifaceted field of public policy, advanced methodologies that use the power of data and computational capacities will play an increasingly significant role in policy-making and governance. A digital revolution is in continuous progress and should be exploited by policymakers and public managers because the field of public policy and administration must adapt to continual changes, and this adaptation requires the development of innovative and efficient tools to cope with the complexity of reality.

References

Allison, G. T. (1971). *Essence of decision: Explaining the Cuban missile crisis*. Boston: Little, Brown.

Epstein, J. M. (2006). Generative Social Science: Studies. In *Agent-Based Computational Modeling*. Princeton, NJ: Princeton University Press.

Janssen, M., Wimmer, M. A. & Deljoo, A. (Eds). (2015). *Policy Practice and Digital Science. Integrating Complex Systems, Social Simulation and Public Administration in Policy Research*. Switzerland: Springer International Publishing.

Klijn, E. H., & Snellen, I. (2009). *Complexity theory and public administration: A critical appraisal*. In G. R. Teisman, A. Buuren, & L. M. Gerrits (Eds.), Managing Complex Governance Systems.

Koliba, C., & Zia, A. (2011). *Theory Testing Using Complex Systems Modeling in Public Administration and Policy Studies: Challenges and Opportunities for a Meta—Theoretical Research Program*. Public Management Research Conference (PMRC). Maxwell School of Citizenship and Public Affairs. Syracuse University.

[3]See on this subject: Daniell, Morton, & Ríos Insua (2016). Policy analysis and policy analytics. Annals of Operations Research January 2016, Volume 236, Issue 1, p. 1–13. See also: Tsoukias, A., Montibeller, G., Lucertini, G. et al. Policy analytics: an agenda for research and practice. *EURO J Decis Process 1*,115–134 (2013).

March, J. G., & Olsen, J. P. (Eds.) (1985). *Ambiguity and Choice in Organizations.* New York: Oxford University Press.

Meek, J. W. (2010). *Editorial: Complexity Theory and administrative learning—Adaptive practices in complex governance systems.* E:CO 2014 vol. 16, no. (1), pp. 1–6. Emergent Publications.

Mitchell, M. (2009). *Complexity: A Guided Tour.* New York: Oxford University Press, Oxford.

Mitleton-Kelly, E. (2003). *Complex Systems and Evolutionary Perspectives on Organizations: The Application of Complexity Theory to Organizations, Bingley.* UK: Emerald Group Publishing Limited.

Solo, A. M. G. (2014). The new interdisciplinary fields of public policy engineering and computational public policy. In A. M. G. Solo (Ed.), *Political Campaigning in the Information Age.* Hershey, PA: IGI Global.

Tsoukias, A., Montibeller, G., Lucertini, G., & Belton, V. (2013). Policy analytics: an agenda for research and practice. *EURO Journal on Decision Processes, 1*(1–2), 115–134.

Zgurovski, M. Z., Zaychenko, Y. P. (2016). *The Fundamentals of Computational Intelligence: System Approach.* Studies in Computational Intelligence, vol. 652. Switzerland: Springer International Publishing.

Cairney, P. & Jones, M. (2015). Kingdon's multiple streams approach: What is the empirical impact of this universal theory? *The Policy Studies Journal, 44.* https://doi.org/10.1111/psj.12111.

Simon, H. A. (1969). *The sciences of the artificial.* M.I.T. Press.

Chapter 9
Conclusion and Future Research

Nothing is more powerful than an idea whose time has come.
Victor Hugo

In this conclusion, we summarize the main steps of this research. We describe the approaches and conclusions of each step and suggest new research avenues.

This research is performed using different theories and approaches. It is motivated by several ambitions. The first objective of this work is to explore the multiple levels of the policy decision process and how these levels are considered by the classical decision-making approaches in public policy.

We reviewed the main classical approaches in public policy decision analysis, namely the rational model, the incremental model, the bureaucratic model, and the Garbage Can Decision Model. Our goal was to examine how these approaches deal with the policy decision process and to identify the level of the decision process examined by each approach. Following this review, we concluded that these analytical models analyze the decision process from a systemic perspective at the macro level, and they are actor-centered and positivist approaches. This review also demonstrated that, for over half a century, analytical approaches in public policy evolved toward a comprehensive exploration of the multiple levels of the decision process. This evolution occurs with the purpose of a more complete understanding of the decision process and decision-making.

The review demonstrates the richness of the decision-making analysis and the interlacing complementary layers of the decision process. The review also shows the epistemological progression of the approaches that start with a positivist and deterministic vision, and moves slowly toward a postpositivist perspective. This perspective includes nonlinear aspects that do not rely on causal links. It also adds a deeper sub systemic dynamic to the decision process that was raised by certain models. However, this sub systemic dynamic remained unexplored by the classical approaches because of the lack of appropriate tools.

The first ambition of this research is, therefore, to determine the sub systemic environment (or micro-level) of the decision process, and to identify its components and its dynamics. The investigation conducted to achieve this objective demonstrates that the sub systemic level of the decision process has characteristics of a complex

A. Guidara, *Policy Decision Modeling with Fuzzy Logic*, Studies in Fuzziness and Soft Computing 405, https://doi.org/10.1007/978-3-030-62628-0_9

system. Such a system is composed of several entities that interact with each other without central control. The dynamics of a complex system generate a global behavior that cannot be predicted from the individual characteristics of its components.

We found that the sub systemic environment of the decision process better corresponds to the post-positivist thinking. Therefore, it requires a paradigmatic shift and appropriate methodologies to deal with. However, the nature and the properties of this sub systemic environment are not known so far.

The challenge we faced was then double. On the one hand, we target the sub systemic environment of the decision process to discover its constituent elements, and their characteristics and interactions. On the other hand, we need to find the appropriate tools to deal with this environment and its dynamics.

Since the sub systemic environment of the decision process has characteristics of a complex system, its inherent dynamics represent the background that generates the global behavior of such a system, in this case, the decision-making circumstances that we defined as the policy decision emergence. This emergence arises at this sub systemic level and it may lead to decision-making.

With these characteristics in mind, our first reflex was to explore the complex systems theory in order to find tools that deal with the sub systemic environment of the decision process. However, complex systems theory is not a homogeneous field, but rather a set of various concepts and methods for dealing with complex phenomena, such as the policy decision emergence defined above.

A review of complex systems theory demonstrates that such dynamic systems require appropriate methodologies to understand their dynamics such as modeling and simulation. Modeling consists in creating a model that represents an approximation of the real system, and simulation consists in experimenting the behavior of the model under different scenarios and conditions to understand its evolution and forecast its outcome. However, the methodology does not provide a technique to determine the structure and configuration of the system. Besides, it cannot address a public policy issue without a theoretical foundation in this field (Morçöl, 2010) that constitutes a groundwork to build the model.

Therefore, we first need to find a method to determine the components of the complex sub systemic environment of the decision process to determine the conceptual model of the decision emergence, and second, we need to find the necessary link to bridge complex systems theory and the field of public policy.

The link is provided by Stacey Matrix that was designed to describe and explain decision-making in management settings. The Stacey matrix is a complexity tool that defines a range of several decision-making zones according to certain criteria, including a zone of complexity. Geyer and Rihani (2010), adapted this matrix to the public policy field and they determined that the complexity zone includes Agenda building (or setting), and Garbage Can Decision Making. These two notions point to the multiple streams theory, developed by Kingdon. Indeed, the multiple streams theory integrates these two notions because it is inspired by the Garbage Can Decision Making Model, and it deals with the agenda-setting.

However, this theory represents an analytical framework dedicated to the study of policy development and change, and it does not constitute a decision-making model to deal with decision emergence. Furthermore, although this theory includes concepts related to the complexity of the decision process, it does not give a suitable technique to operationalize this process.

Kingdon's theory outlines three streams as the basis of agenda-setting and policy development: the problem stream, the solution stream, and the politics stream. We noticed that these notions are not only global aspects in the context of policy-making, but they are also vague and uncertain. Any attempt to characterize these streams represents a risk of approximation and linearization that neglects the complexity of the decision process. Moreover, such an attempt constitutes a systemic approach that does not consider sub systemic aspects. We concluded that multiple streams theory, as formulated by Kingdon, does not meet the needs of our research project. Nevertheless, it establishes a link between complex systems theory and the field of public policy through Stacey Matrix. Therefore, it constitutes the wanted link to bridge complex system theory and public policy and it also provides the main variables (the streams) of the system that characterize any decision-making process. Indeed, for a decision to be made and a policy to be developed, there must be a problem, a solution, and a political will. Consequently, we focused on an in-depth review of multiple streams theory and related academic works that provide details about the variables of the system, and especially about the sub systemic components of these variables.

This review allows us to identify the sub systemic factors within each stream or variable. Furthermore, these factors can be operationalized and assessed.

We retained these factors as the sub systemic components of the decision process that generates the decision-making circumstances, namely the decision emergence. Indeed, the dynamics of this micro-level environment and the interactions among its components determine the evolution and the behavior of the streams, as concluded from our review of the multiple streams theory.

Our conceptual model is therefore made-up of the sub systemic factors of each stream, identified as micro-level components, and the variables (streams) as middle-level components of the decision emergence system. The dynamics of the system generate decision circumstances. This assumption is inspired by the notion of streams coupling developed by Kingdon. We called our model the Policy Decision Emergence Simulation Model (PODESIM).

However, while the multiple streams theory allowed us to identify the factors and the variables of the system underlying the decision emergence, it does not address the dynamics of these components nor their individual influence on the global evolution of the system. To get around this limitation, it was necessary to analyze further the properties of the components to find the appropriate tools that determine their dynamics. Besides, it was also necessary to find an appropriate computing application for the simulation.

Following a careful analysis, we determined that all components of the model represent approximate and vague concepts. Although some of them are quantifiable, their assessment is subject to interpretation and their influence depends on human reasoning and estimation, which are subjective by nature. This aspect brought an additional challenge to this research because we needed to consider in our model the approximate nature of information and human reasoning which are characterized by vagueness, uncertainty, and ambiguity.

For example, referring to a problem, one of the variables of the system, as urgent is a vague and imprecise description because it depends on the context, on the individual reasoning or groupthink, and other tangible and intangible factors.

Nevertheless, all components of the conceptual model can be described by linguistic variables that determine their scale or magnitude, even the quantifiable ones. For example, having an unemployment rate of 10% can be considered as a high rate in one place but a medium or even a low rate in other places. This assessment depends on the context and other criteria.

Based on this deduction, we identified fuzzy logic as an appropriate theory to deal with vagueness, imprecision, approximate information, symbolic reasoning, and partial truth that characterize the components of the model PODESIM. This step represents the integration of a third scientific approach in this multidisciplinary research, in addition to a combination of multiple streams theory and complex system theory.

Fuzzy logic is a mathematical discipline and an artificial intelligence application. It gave rise to fuzzy sets theory, a research approach that deals with subjective and imprecise data that can be described by linguistic variables. The fuzzy sets are classes of elements whose boundaries are gradual in which the information is expressed by possibility distributions. Fuzzy sets elements are represented by membership functions that illustrate the degree of fuzziness or degree of truth of their elements.

The progression of the fuzzy sets theory led to the development of fuzzy inference systems and fuzzy inference rules (decision heuristics). This development facilitated modeling and simulation that transform fuzzy inputs into outputs through an inference algorithm and a set of inference rules. This technique consists of building models with fuzzy inference systems as building blocks to compute the imprecise and uncertain properties of the components of a system.

Because of this uncertainty and imprecision, the components of the system are designated by partial scales intervals, which are described by linguistic variables (e.g., small, medium, large). The set of intervals represents the full range of the component's possible values and is referred to as the universe of discourse.

The technique then consists of associating decision heuristics in the form of inference rules, based on knowledge and expertise in the field. These rules are heuristics that assume the evolution of the system according to a set of possible scenarios. The inference rules are formulated using logical operators AND, OR, NOT to carry out the association of the conditions attributed to the different scenarios. Ross (2010) argues that [...] the power of fuzzy nonlinear simulation is manifested in modeling nonlinear systems whose behavior we can express in the form of input-output data-tuples, or in the form of linguistic rules of knowledge, and whose exact nonlinear specification we do not know (p. 265).

The choice of fuzzy logic theory met our need to build the computational model and carry out the simulation with two different configurations to experiment with the model. Each of the two configurations includes a type of membership function describing the partial ranges of values, combined with two defuzzification methods, in order to determine which configuration produces the most reliable results that are corroborated by the facts and events. The results given by simulation allowed us to determine which configuration gives the most credible results.

This "experimentation" is followed by validation through an empirical case study. Validation is required in any modeling and simulation process to verify that the model is performing as expected in real situations and that the outputs are in line with their design objectives. Validation also allows to fine-tune the model if necessary.

For this step, we selected the Cuban missile crisis case. Our goal is to simulate the decision emergence that generates the decision-making of the American administration during the crisis period. This crisis lasted officially from October 15 to October 28, 1962, and we simulated the decision emergence for each day of this period.

The choice of this empirical case is justified by several aspects. First, the fact that this crisis still stimulates the interest of researchers because it has not yet revealed all its secrets and can still shed light that could help decision-makers (Dobbs, 2008). Second, because previous analyses of this crisis have been carried out using conventional analytical approaches that deal with decision making at the systemic level and neglect unpredictable elements (Dobbs, 2008).

Finally, the most important element for the study of this case is above all the availability of detailed information from primary sources concerning each day of the crisis. The most valuable and reliable source was the transcripts of the meetings of the crisis unit (ExCom) set up by President Kennedy at the beginning of the missile crisis. These meeting were secretly recorded by Kennedy without informing the members of the ExCom, which makes them an authentic primary source that did not undergo any modification or interpretation.

For this empirical case study, we first proceeded with a data collection and coding using NVivo® software, then we carried out simulations with Matlab® Fuzzy Logic Toolbox™ dedicated to fuzzy inference systems. The objective of the simulation is to identify the specific factors that contribute to generate decision-making circumstances. These factors would be the levers of decision emergence.

The results given by simulation demonstrated that the PODESIM model works as expected and we could validate with a certain level of confidence that the model's fuzzy inference systems are consistent and reliable for this case study. Furthermore, the results also confirmed that the phenomenon of decision emergence, as a global behavior of a complex system, is not a result of cause-effect relationships, and it cannot be predicted from the individual state of the system's components.

The results also confirmed that no correlation or linear causal link exists among the components at the sub systemic level (the factors) and the overall systemic behavior at the macro level, the decision emergence.

Regardless of the scale or the value of factors and variables, these setting do not individually direct the evolution of the system and do not predict the decision emergence. The decision emergence is a complex system in which the state of the components cannot predetermine their impact on global behavior. It is the state of all components and their interactions that generate systemic behavior.

Finally, the findings and the validation also confirmed that the methodology chosen for this research is an appropriate one and that PODESIM is a reliable model for the simulation of policy decision emergence as a complex system using fuzzy logic.

In conclusion, the PODESIM model could answer our research question. It allowed us to determine the sub systemic components of the decision process, and to identify the levers and the scale of the decision emergence, for the case study. This work also highlights characteristics and details of the decision process that are not addressed by the analytical approaches, which makes PODESIM a new appropriate tool for decision diagnostics, monitoring, and forecasting that goes beyond classical policy decision analysis.

However, one empirical case study is not sufficient to conclude that the model is robust and stable. We raised limitations and uncertainties of the model in our analysis of results, but these limitations are related to the case study. Therefore, like any simulation model, PODESIM requires further experimentation and more case studies to be improved and consolidated.

Because it represents a complex system, PODESIM is an open multi-level environment that can integrate additional fuzzy inference systems and additional factors and subfactors. It can easily adapt to different environments with different properties and levels, but it requires further development.

Therefore, we suggest the following non-exhaustive list of directions for future research to advance the PODESIM model and expand its application area.

Besides the fuzzy inference systems that already constitute the backbone of PODESIM, it would be useful and stimulating to add other components that represent other aspects of the policy decision process not yet considered. For example, identifying and adding components of the international political environment, that influence the domestic decision process, in the form of additional factors to the existing fuzzy inference systems (FIS) or in the form of variables as additional fuzzy inference systems. The Advocacy Coalition Framework (Sabatier, 2007) and the notions of political entrepreneurs (Kingdon, 2014) for example are an inspiring source to enrich the model. However, this represents an additional challenge to formulate appropriate fuzzy inference rules that require expertise in the field of international politics.

Moreover, we propose to identify additional factors and subfactors to existing fuzzy inference systems of PODESIM. Adding a lower level of sub-factors can refine the model and perhaps learn, through the simulation, new details that underly the decision emergence. Since PODESIM is an open system, it is possible to add to its structure as many levels as necessary as long as the expertise in the field can manage the details of multiple levels and rules.

Advanced data collection and processing techniques can contribute to identifying new factors and subfactors, and help to determine consistent membership functions and reliable inference rules. Also, other methodologies such as agent-based modeling (ABM) and artificial neural networks (ANN) can add tremendous details to the factors and their dynamics.

Any other techniques that generate time series for the simulation could also be useful to enhance the simulation of the model. Moreover, the integration of specific time series to the existing factors helps to determine more consistent membership functions which result in more accurate findings. Appropriate time series can incorporate large volumes of data and allow an automatic simulation. Automation offers the possibility of performing real-time simulation and monitoring of a system, which may be useful to monitor the implementation of policies and to carry out a real-time assessment of decisions and policies.

Regarding the inference rules or decision heuristics which are based on the expertise and judgment, we propose an integration of rules from classical decision approaches when it is suitable or required. For example, the rational model or the incremental model (Lindblom, 1959) can inspire some rules in certain cases. Allison's findings (1971) can also suggest specific inference rules associated with the organizational units and with bureaucratic bargaining. But these propositions could represent a huge challenge.

From another perspective, we believe that inference rules should also consider cognitive decision heuristics (Pomerol and Adam, as cited in Phillips-Wren, Ichalkaranje, and Jain, 2008) that go beyond rational thinking. Political decisions are sometimes influenced by individual values and beliefs, and sometimes by emotional aspects such as a potential of reward or a desire for revenge. This, of course, is another enormous challenge that deserves a starting point.

Finally, advanced techniques can also generate fuzzy inference rules in certain cases. Evolutionary algorithm (EA), such as genetic algorithm, is an example of these techniques.[1] Evolutionary algorithms are programs that analyze large amounts of data from a variety of sources to extract information that would otherwise be inaccessible. It is this property that gives these algorithms the ability to generate fuzzy inference rules.

The propositions above show the high potential of PODESIM to advance the field of computational decision-making in public policy thanks to its flexibility and adaptability. The findings of this research and the suggested future avenues position the PODESIM model at the core of the policy decision-making modeling and simulation, and the governance structure.

In the following Fig. 9.1, we summarize the purpose of modeling and simulation in the political and organizational environment. These techniques require further research on digital transformation [2] in the field of public policy and public administration.

[1]See: Baron, Achiche, Balazinski, (2001).

[2]Digital transformation refers to the processes of integrating digital technologies into all organizational activities and spheres.

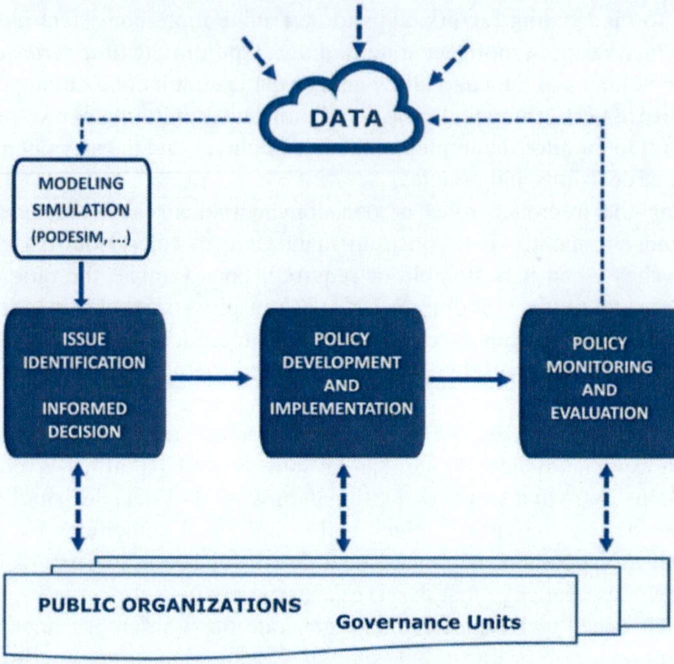

Fig. 9.1 PODESIM model in the policy and administration structure

Modeling and Simulation for Public Policies and Governance

Modeling and simulation contribute to the ongoing digital transformation of public policies. This digital transformation is a big challenge because it depends on three intertwined aspects: Data collection and management, advanced computational methodologies, and reliable political and organizational structure.

The methodology developed in this research contributes to the paradigm shift from the field of classical policy analysis to the Computational Policy field. A new research program that aims to address contemporary challenges in public policy and administration through the development of advanced methodologies and computing tools, and to supplement the digital transformation of policies.

This work contributes to this research program that relies strongly on two important resources: information and expertise. It opens new paths for the integration of methods and tools of data analysis, modeling and simulation, and computing techniques. This integration is necessary for the development of strategic and efficient policies in this era of a new digital revolution.

References

Allison, G. T. (1971). *Essence of decision: Explaining the Cuban missile crisis*. Boston: Little, Brown.

Baron, L., Achiche, S., Balazinski, M. (2001). Fuzzy decision support system knowledge base generation using a genetic algorithm. *International Journal of Approximate Reasoning, 28*, 125–148. Elsevier.

Dobbs, M. (2008). *One minute to midnight: Kennedy, Khrushchev, and Castro on the brink of nuclear war*. New York: Vintage Books.

Geyer, R., & Rihani, S. (2010). *Complexity and public policy: A new approach to twenty-first century politics. Policy and society*. Abingdon: Routledge.

Kingdon, J. (2014). *Agendas, alternatives and public policies* (2nd ed.). Essex: Pearson Education Limited.

Lindblom, C. E. (1959). The science of muddling through. *Public Administration Review, 19*(2), 79–88.

Morçöl, G. (2010). Issues in reconceptualizing public policy from the perspective of complexity theory. *Emergence: Complexity and Organization, 12*(1).

Phillips-Wren, G., Ichalkaranje, N., & Jain, L. C. (Eds.). (2008). *Intelligent decision making: An AI-based approach*. Berlin, Heidelberg: Springer-Verlag.

Ross, T. J. (2010). *Fuzzy logic. With engineering applications* (3rd Edn.). Wiley.

Sabatier, P. A. (2007). *Theories of the policy process*. Boulder, Colorado: Westview Press.

Uncited References

Axelrod, R. (1997a). Advancing the art of simulation in the social sciences. In D. R. Conte, R. Hegselmann, & P. Terna (Dir.), *Simulating social phenomena* (Vol. 456), Lecture notes in economics and mathematical systems (pp. 21–40). Berlin: Springer.

Axelrod, R., & Cohen, M. D. (2000). *Harnessing complexity: Organizational implications of a scientific frontier*. New York: The Free Press.

Axelrod, R., & Tesfatsion, L. (2006). A guide for newcomers to agent-based modeling in the social sciences. Appendix A. In D. L. Tesfatsion & K. L. Judd (Dir.) *Handbook of Computational Economics* (Vol. 2): Agent-based computational economics, Handbooks in economics series. Amsterdam, the Netherlands: Elsevier North-Holland.

Axtell, R. (2000). Why agents? On the varied motivations for agent computing in the social sciences. Working Paper No. 17, *Center on Social and Economic Dynamics*. The Brookings Institution.

Bělohlávek, R., Billari, J. W., & Klir, G. J. (2017). *Fuzzy logic and mathematics: A historical perspective*. New York: Oxford University Press.

Edmonds, B. (2010). Bootstrapping knowledge about social phenomena using simulation models. *Journal of Artificial Societies and Social Simulation*, 13(1), 8. Available at http://jasss.soc.surrey.ac.uk/13/1/8.html.

Epstein, J. M. (2008). Why Model? *Journal of Artificial Societies and Social Simulation, 11*(4)12. Available at http://jasss.soc.surrey.ac.uk/11/4/12.html.

Grimm, V., Berger, U., Bastiansen, F., Eliassen, S., Ginot, V., …, DeAngelis, D. L. (2006). A standard protocol for describing individual-based and agent-based models. *Ecological Modelling, 198*(1–2), 115–126.

Grimm, V., Berger, U., DeAngelis, D. L., Polhill, J. G., Giske, J., & Railsback, S. F. (2010). The ODD protocol: A review and first update. *Ecological Modelling, 221*(23), 2760–2768.

Harrison, N. E. (Dir.). (2006). *Complexity in world politics, concepts and methods of a new paradigm*. Albany: State University of New York Press.

Helbing, D. (Dir.) (2012). *Social self-organization. Agent-based simulations and experiments to study emergent social behavior*. Berlin, Heidelberg: Springer.

Holland, J. H. (1998). *Emergence. From Chaos to Order*. Cambridge, Massachusetts. Perseus Books.

Jervis, R. (1997). *System Effects. Complexity in Political and Social Life*. New Jersey: Princeton University Press.

John, P. (2012). *Analyzing Public Policy* (2nd ed.). London and New York. Routledge.

Johnson, L. (2016). *Agent-Based Model Basics. A Guidebook & Checklist for Policy Researchers*. Washington. Westphalia Press.

Kegley, C. W., & Blanton, S. L. (2011-2012). *World politics: Trends and transformations*. Cengage Learning.

A. Guidara, *Policy Decision Modeling with Fuzzy Logic*, Studies in Fuzziness and Soft Computing 405, https://doi.org/10.1007/978-3-030-62628-0

Keller, J. M., Liu, D., & Fogel, D. B. (2016). *Fundamentals of computational intelligence. Neural networks, fuzzy systems, and evolutionary computation*. Hoboken, New Jersey: Wiley.

Lempert, R. (2002). Agent-based modeling as organizational and public policy simulators. *Proceedings of the National Academy of Sciences, 99*(3), 7195–7196.

Lincoln, Y. S., & Guba, E. G. (1985). *Naturalistic Inquiry*. Newbury Park, CA: Sage Publications.

Mamdani, E. H. & Assilian, S. (1975). An experiment in linguistic synthesis with a fuzzy logic controller. *International Journal of Man-Machine Studies, 7*(1), 1–13.

Marquis, P., Papini, O. & Prade, H. (Coord.) (2014). *L'intelligence artificielle : frontières et applications*. Toulouse: Cépaduès Éditions.

Morçöl, G. (2003). Complexity and public administration. Public administration theory. In *Network Annual Conference*, Anchorage, Alaska.

Railsback, S. F., Lytinen, S. L., & Jackson, S. K. (2006). Agent-based simulation platforms: Review and development recommendations. *Simulation, 82*(9), 609–623.

Rhodes, D. H., & Ross, A. M. (2010). Five aspects of engineering complex systems emerging constructs and methods. In *2010 IEEE International Systems Conference*, San Diego, CA (pp. 190–195).

Rutkowski, L. (2008). *Computational intelligence. Methods and techniques*. Berlin Heidelberg: Springer.

Simon (1997), H. A. (1955). A behavioral model of rational choice. *Quarterly Journal of Economics, 69*, 99–118.

Stern, S. M. (2012). *The Cuban missile crisis in American memory. Myths versus reality*. Stanford University Press.

Taylor, S. J. E., (Dir.). (2014). *Agent-based modeling and simulation*. Palgrave Macmillan.

Wilensky, U., & Rand, W. (2015). *An introduction to agent-based modeling. Modeling natural, social, and engineered complex systems with Netlogo*. Massachusetts Institute of Technology.

Yager, R. R., & Zadeh, L. A. (Eds.) (1992). *An introduction to fuzzy logic applications in intelligent systems*. New York: Springer Science+Business Media.

Zadeh, L. A. (1973). Outline of a new approach to the analysis of complex systems and decision processes. *IEEE Transactions on Systems, Man, and Cybernetics, 3*(1), 28–44.

Zadeh, L. A. (1979). *A Theory of approximate reasoning*. In J. Hayes, D. Michie, & L. Mikulich (Eds). *Machine intelligence* (pp. 149–194). New York: Halstead Press.